After
the
Earth
Quakes

After the Earth Quakes

Elastic Rebound on an Urban Planet

Susan Elizabeth Hough
and Roger G. Bilham

OXFORD
UNIVERSITY PRESS

2006

OXFORD
UNIVERSITY PRESS

Oxford University Press, Inc., publishes works that further
Oxford University's objective of excellence
in research, scholarship, and education.

Oxford   New York
Auckland   Cape Town   Dar es Salaam   Hong Kong   Karachi
Kuala Lumpur   Madrid   Melbourne   Mexico City   Nairobi
New Delhi   Shanghai   Taipei   Toronto

With offices in
Argentina   Austria   Brazil   Chile   Czech Republic   France   Greece
Guatemala   Hungary   Italy   Japan   Poland   Portugal   Singapore
South Korea   Switzerland   Thailand   Turkey   Ukraine   Vietnam

Published by Oxford University Press, Inc.
198 Madison Avenue, New York, New York 10016

www.oup.com

Oxford is a registered trademark of Oxford University Press

Library of Congress Cataloging-in-Publication Data
Hough, Susan Elizabeth, 1961–
After the Earth quakes : elastic rebound on an Urban planet /
by Susan Elizabeth Hough and Roger G. Bilham.
     p.   cm.
Includes bibliographical references and index.
ISBN-13 978-0-19-517913-2
ISBN 0-19-517913-7
1. Elastic rebound theory. 2. Earthquakes. I. Bilham, Roger G. II. Title.
QE539.2.E42H67 2005
551.22—dc22   2004018347

9 8 7 6 5 4 3 2 1

Printed in the United States of America
on acid-free paper

For
Nick Ambraseys,
with appreciation

Contents

After
the
Earth
Quakes

1

Impacts and Reverberations

Human society is organized for a stable earth; its whole
machinery supposes that, while the other familiar elements
of air and water are fluctuating and untrustworthy, the earth
affords a foundation which is firm. Now and then this implied
compact with nature is broken, and the ground trembles
beneath our feet. At such times we feel a painful sense of
shipwrecked confidence; we learn how very precious to us
was that trust in the earth which we gave without question.
—N. S. Shaler, "The Stability of the Earth,"
Scribner's Magazine (March 1887): 259

Earthquakes and their attendant phenomena rank among the most terrifying natural disasters faced by mankind. Out of a clear blue sky—or worse,
a jet-black one—comes shaking strong enough to hurl furniture across the
room, human bodies out of bed, and entire houses off their foundations. Individuals who experience the full brunt of the planet's strongest convulsions
often later describe the single thought that echoed in their minds during the
tumult: I am going to die. When the dust settles, the immediate aftermath of
an earthquake in an urbanized society can be profound. Phone service and
water supplies can be disrupted for days, fires can erupt, and even a small
number of overpass collapses can impede rescue operations and snarl traffic

for months. On an increasingly urban planet, millions of people have positioned themselves directly in harm's way. Global settlement patterns have in all too many cases resulted in enormous concentrations of humanity in some of the planet's most dangerous earthquake zones.

On the holiday Sunday morning of December 26, 2004, citizens and tourists in countries around the rim of the Indian Ocean were at work and at play when an enormous M9 (magnitude 9.0) earthquake suddenly unleashed a torrent of water several times larger than the volume of the Great Salt Lake. The world then watched with horror as events unfolded: a death toll that climbed toward 300,000 that was accompanied by unimaginable, and seemingly insurmountable, devastation to hundreds of towns and cities.

For scientists involved with earthquake hazards research in that part of the world, the images were doubly wrenching: the hazard from large global earthquakes has been recognized for decades. Located mostly offshore, the 2004 Sumatra quake unleashed its destructive fury primarily in the sea. The next great earthquake to affect Asia might well be inland, perhaps along the Himalayan front or in central China. The toll from this next great quake, whether it is as large as 9 in magnitude or "merely" a low M8 temblor, could be far worse: little if any tsunami damage but potentially catastrophic damage from strong shaking under one or more population centers.

The odds are good that Asia will be as unprepared for the next great quake as it was for the last one.

The images will be every bit as horrifying.

Scientists will feel every bit as impotent.

In increasingly restrictive economic times, proposals to implement even rudimentary warning systems and other mitigation efforts in developing parts of the world invariably land like lead balloons. Such efforts can be enormously cost-effective: a tsunami alert system could have been installed around the rim of the Indian Ocean for perhaps $20 million—about as much as the United States spent on the 2004 Iraq war effort every three hours. Worse still, many lives lost on distant shorelines could have been saved for a cost of a few hundred dollars per village by combining existing warning sirens with modest public education about potential dangers from the sea. For scientists who understand best what is at stake, the frustration and helplessness could scarcely be more personal.

As scientists we do, however, recognize that earthquakes are only one concern among many for citizens of developing nations. When basic needs such as adequate food and safe water are unmet, earthquake safety can seem like a luxury. In virtually every corner of the world, earthquakes are, moreover, scarcely the only natural disaster of concern.

Worse yet, some of the more horrific disasters in recent years, in wealthy as well as poorer countries, have had nothing to do with the powerful forces of nature. Shock waves reverberated around the planet after the World Trade Center towers collapsed on the horrific morning of September 11, 2001. Even for those who had seen the catastrophic damage and conflagration caused by the two airplane impacts, it seemed so wildly improbable—so impossible— that such tall and proud structures might actually fall down. When they fell, the earth itself registered the shock. Thousands of tons of falling steel and debris literally generated waves within the ground, exactly the same kind of waves generated by earthquakes. If the figurative impact of 9/11 was enormous, the literal impact on the earth was perhaps surprisingly modest. Seismometers operated by the Lamont–Doherty Earth Observatory registered tiny earthquakes (M0.9 and M0.7, respectively) generated when the two airplanes crashed into the towers. When the buildings fell, their reverberations were equivalent to larger but still minor earthquakes, with magnitudes of 2.1 and 2.3.

Natural earthquakes with similarly humble magnitudes do occur in New York City from time to time. In this part of the world the earth's crust comprises old and cold rocks through which earthquake waves travel efficiently. It is by no means unheard of for even a lowly M2 earthquake to be felt in this type of geologic environment. When an M2.5 temblor struck the Upper East Side of Manhattan on the morning of January 17, 2001, a U.S. Geological Survey Web site received over 100 reports from individuals who felt the shock— from distances as far as 30 miles from the Upper East Side. It is thus by no means unheard of for the news media to respond with interest when even a tiny temblor shakes things up.

The modest seismic reverberations on the morning of September 11 appeared to belie the extent of the disaster from which they had sprung. At ground zero, meanwhile, the collapsing towers bore little resemblance to a small earthquake but every resemblance to another geological phenomenon: the so-called pyroclastic flow—a fast-moving cascade of hot dust and debris—generated by certain kinds of volcanic eruptions. The force of the cascading avalanche claimed victims outside of the towers as well as those trapped inside, and left shell-shocked survivors caked in layers of ghostly gray dust. It was, indeed, as if a volcano had erupted, violently and without warning, in the heart of Lower Manhattan, the towers themselves reduced to a shell as meager and flimsy as the one left after Mt. Saint Helens exploded in Washington State in 1980.

Much has been written about the tumultuous events of September 11; it will be left to history books yet to come to assess the full legacy of this day. One suspects these books will not all agree. The short-term legacy of 9/11 was, however, as dramatic as it was indisputable: heroism on the part of those who were in a position to act, be it on a hijacked airplane or in a New York City's firefighter's uniform, and patriotism on the part of everyone else. The Stars and Stripes appeared everywhere. From Lower Manhattan to neighborhoods in California where shop signs are in Mandarin, patriotism was suddenly on display where patriotism seemingly hadn't existed before. On September 12, the front-page headline of *Le Monde* proclaimed, "We are all Americans."

This is not a book about September 11, nor is it about politics or world events in the usual sense of the phrase. It is a book about planetary events: specifically, the large and sometimes devastating earthquakes, such as the 2004 Sumatra quake, that strike our planet, and the impact of these temblors on individuals and societies. Natural disasters are, of course, both like and unlike unnatural disasters such as the 9/11 terrorist attacks. By now, the occurrence of most large earthquakes is not a shock (so to speak) in the scheme of things, but the occurrence of a rare large earthquake in a relatively quiet part of the planet can rock (so to speak) sensibilities nearly as much as an out-of-the-blue act of terrorism. Imagine for a second that an M7.2 earthquake had struck Lower Manhattan at the start of the workday on September 11, 2001. Had such an event happened in 2001, there can be little doubt that the consequences would have been dire; like many U.S. cities in relatively inactive geologic regions, New York City is not well prepared for the large earthquakes that might someday strike. Earth scientists would not have been stunned by such an event; for the public it would have been another matter.

In other parts of the country and the world, occasional large earthquakes are not only a part of life but also an integral part of the cultural fabric. Californians live in earthquake country. It should surprise no modern Californians over the age of eight when the ground springs to life with no warning. Even in historic times earthquakes could not have been considered a surprise: the very earliest records kept by explorers of European descent include mention of earthquakes both felt and reported by California's native peoples. In 1769 the Gaspar de Portola expedition experienced a series of strong earthquakes as it made its way through what is now Orange County. What we know today as the Santa Ana River was christened in 1769 as the Santa Ana de los Temblores. Thus California has been known as earthquake country since literally the earliest days that written records were kept in the state. Forewarned is forearmed, as they say, but even in earthquake country one can never be truly prepared for that moment when the terra firma abruptly and rudely ceases to be firm.

Large earthquakes, be they anticipated in a certain area or not, challenge our sensibilities in a manner not unlike that of the 9/11 terrorist attacks. As N. S. Shaler wrote more than a century ago, unleashed planetary forces dissolve in a heartbeat the very essence of stability on which a rational, orderly human society is built.[1] The metaphors of daily life speak volumes: ground truth, bedrock values, terra firma, rock-solid. The earth revolves around the sun, but the earth beneath our feet is not supposed to move. When it does, the impact goes far beyond the immediate and obvious consequences, catastrophic as they may be. People don't like earthquakes. In the United States, people don't like earthquakes in spite of the very low risk that a person will ever die, or be seriously injured, in a major temblor. People don't like earthquakes because of the irreducible, nonnegotiable elements of unpredictability, surprise, and terror.

Individual earthquakes cannot be predicted, but the long-term rates of earthquakes can now be forecast with considerable accuracy, particularly in the planet's seismically active plate boundary regions, such as Italy, Japan, Turkey, Mexico, and California. In these areas, large earthquakes are now known to occur with predictable, if imperfect, regularity. This predictability is, however, limited to a long-term sense: large earthquakes such as the 1906 San Francisco temblor will recur on average every 200 years; great earthquakes will strike near Tokyo about twice as often, on average. (These values represent only averages: large earthquakes do not recur like clockwork.) Many of the most powerful civilizations of today and yesteryear have, with seemingly tragic irony, grown up directly atop these very regions. People don't like earthquakes, and yet, over and over again, people choose to live in areas susceptible to earthquakes.

This coincidence—great civilizations and active plate boundaries—is in fact less senseless than it might appear at first blush. Apart from the occasional terrifying and catastrophic earthquake, plate boundary zones often offer far more than their fair share of geographical amenities. Many plate boundaries run directly or nearly directly along coastlines, which have beckoned to human settlements for obvious reasons since the dawn of time. (Even when coastlines are not along active plate boundaries, such areas, as the world has witnessed, can still be horribly vulnerable to tsunami damage from relatively distant quakes.) Active plate boundaries create mountains that can buffer local climates. Mountains also tend to impede the migration of populations: their rugged terrain is generally inhospitable to settlement, and their leeward sides are often hot and dry. Where oceanic crust sinks beneath continents along coastlines, the process of subduction builds mountains not far inland (figure 1.1), creating a narrow strip of hospitable coastline along which populations invariably congregate. Chile is a long and narrow country, but the overwhelming majority of the country's population lives in an even narrower band of longitude sandwiched between the Pacific Ocean to the west and the Andes Mountains to the east. In Japan, active plate boundary processes give rise to (literally) the islands on which the country is built.

Indeed, if the earth were not a dynamic planet, there would be no life in the first place—at least, not the life that has evolved here over a billion-year time scale. In the mid-19th century, the physicist William Thomson, later Lord Kelvin, argued that the planet could not possibly be as old as 1 million years, an argument he based on the current temperature of the earth and the rates at which a planetary body was expected to cool. By his calculations, the earth would have been far too hot to support life as little as a million years in the past. These results were most vexing to another great scientist of the day, Charles Darwin, who crudely estimated the age of the earth to be on the order of several hundred million years—this based largely on his understanding of geology. Darwin's estimate flew in the face not only of Lord Kelvin's calculation but also of the sensibilities of those still inclined to believe Bishop James Ussher's 17th-century estimate, based on biblical genealogy, that the earth was a mere 6,000 years old. Today, of course, scientists know the earth to be far older than even Darwin guessed: 4.5 billion years, give or take a few hundred million. Kelvin's seemingly impeccable calculations went awry because at the time he performed his calculations, scientists had no understanding of radioactivity—and hence no way of knowing that, by virtue of the especially long and prodigious process of radioactive decay, the planet has cooled far more slowly than it would have if only chemical and gravitational forces were at work. Only as the 19th century drew to a close did scientists have their first inkling that very small atoms could generate very great amounts of energy. Radioactivity has kept the planet toasty warm for a

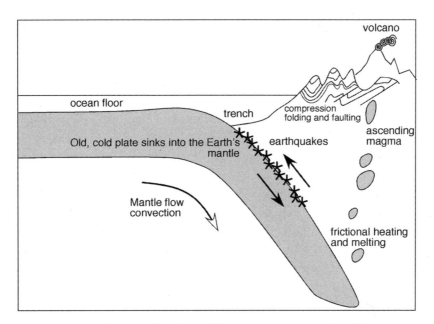

FIGURE 1.1. A cross section through a subduction zone. Compressional forces across the zone commonly give rise to fold mountains. Closer to the coast one finds a chain of mountains of volcanic origin. As the subducting ocean crust sinks, the more volatile material heats up and becomes buoyant, rising through the crust and creating conditions in which magma can form.

very long time, a long enough time for life to develop from the primordial soup and evolve into sentient and intelligent beings. Radioactivity also provides the fuel that drives convection deep within the earth; this convection in turn drives the motion of the planet's large tectonic plates.

It is perhaps unfortunate that our fear of earthquakes and volcanoes disinclines us to appreciate the fact that human beings live on the most fascinating planet in the solar system. Neither a giant ball of gas nor an inert ball of rock, the earth is a planet that is almost entirely solid, yet still dynamic; alive in complex and wonderful ways. And because it is alive, we are alive. That the planet is able to terrify us should not keep it from amazing us as well.

But the planet does terrify us with its occasional, rudely abrupt upheavals. Mankind has come of age on a dynamic planet. Our understanding of earthquakes and volcanic eruptions has progressed from the realm of religion and superstition to the realm of modern (one likes to think rational) science. But what about our response to earthquakes? Anyone who experiences strong shaking, or even pretty strong shaking, cannot help but think that the initial response to an earthquake—the instantaneous, raw, heart-pounding sense of terror—is probably little different for 21st-century humans than it was during caveman days. In a broader sense, however, presumably our response has evolved as scientific understanding has grown and societies have become increasingly complex and interconnected.

What is the impact of devastating earthquakes? The question at first blush appears nonsensical: the impact of devastating earthquakes is, of course, devastation. (As a 21st-century 13-year-old would say, *duh.*) Early images from around the Indian Ocean in late 2004 left the world wondering how the region could possibly recover. But like the collapse of the World Trade Center towers, large earthquakes reverberate in complex and sometimes surprising ways. This book is about those reverberations. It is about the great earthquakes that have occurred throughout recent and not-so-recent history and the impact that these natural disasters have had on individuals and societies—and on the development of scientific understanding of earthquakes as a natural phenomenon. It is about the fundamental element of response that springs from traits that are, and have been throughout time, innately and irreducibly human. It is also about the more complex machinery of societal response: how this type of response has changed throughout mankind's history, yet has at its roots the same elements that define man's innately human response to disasters of all shapes and sizes. And at the end of the day, any book about the impact of large earthquakes must inevitably look forward to consider the earthquakes that in the future will strike our increasingly urbanized planet.

Earthquakes of Yesteryear

To understand the impact of large earthquakes on both societies and the development of scientific thought, one must first understand the earthquakes themselves, how they were understood by scientists of the day, and how our understanding of older earthquakes has developed as the field of seismology has progressed. As a field of inquiry, seismology is markedly young, largely because among the sciences it is somewhat unusual: in essence, seismology involves the investigation of phenomena that are fleeting, unpredictable, and infrequent. One does not scrutinize an earthquake the way one scrutinizes a dinosaur fossil, a Galapagos tortoise, or a heavenly body. At best one catches an earthquake in the act, preserving not the beast itself but some signature of the tracks it created during its brief lifetime.

Invariably and inevitably, the field of seismology leaps forward when large earthquakes strike. Any one earthquake can take the field only so far, the size of the step determined by the state of understanding at the time as well as the degree of sophistication of the monitoring instruments that capture an earthquake's tracks.

Some earthquakes have provided the impetus for wholly new ideas; others have helped crystallize ideas that were already swirling among the best minds of the day. The story of any important historic earthquake—and those of some important modern earthquakes—invariably represents a voyage of scientific discovery. These stories are about the formulation of scientific ideas, but they are also rich in color associated with both historic context and the remarkable individuals whose intellect and vision did so much to build the field of earthquake science as we know it today.

Perhaps not unlike unhappy families (in Tolstoy's estimation), every historic earthquake is important in its own way. Clearly the location, magnitude, and detailed effects of any large earthquake are unique. Important historic earthquakes, however, tend to be important for different reasons. This book is organized largely as a tour through time: each chapter focuses on at least one important earthquake, its societal impact, and its role in the development of modern earthquake science.

As one might imagine, both earthquakes and volcanoes have figured prominently in human history since the dawn of recorded time. Aristotle wrote of subterranean winds, or vapors, that caused earthquakes ranging from "shakers" to "howlers."[2] Pliny the Elder—a naturalist with wide-ranging interests—witnessed, and perished in, the A.D. 79 eruption of Mount Vesuvius. Observing fossil shells on mountaintops, the great 17th-century experimentalist Robert Hooke deduced that the mountains must have been formerly underwater, and that "they themselves most probably seem to have been the effects of some very great earthquake."[3] But also in the mid-17th century, Astronomer Royal John Flamsteed observed that temblors were felt more strongly in the upper floors of buildings and concluded, reasonably but quite wrongly, that earthquakes were caused by explosions in the air (figure 1.2).

A great earthquake offshore of Portugal in 1755 wreaked havoc in the city of Lisbon and sent long-lasting ripples through the European scientific and academic communities, providing impetus for a more scientific consideration of earthquakes than had previously been the norm. The impact of the Lisbon earthquake was that much greater for having followed by just five years a series of widely felt earthquakes that rocked England and captured

FIGURE 1.2. Because the Catania 1692 earthquake was felt by mariners and felt more strongly in the upper floors of tall buildings than at ground levels, John Flamsteed (who sketched this on May 1, 1693) suggested that earthquakes were caused by explosions in the air (E) that rocked buildings X and T outward toward distant points V and W. Flamsteed's quaint caption begins, "*Conceive an Explofion in our Air at E, betwixt the Buildings T and X over the Ship. . .*" (From C. Chamberlaine, "A Letter Concerning Earthquakes Written in the Year 1693 by the Astronomer Mr. John Flamsteed to a Gentleman Residing in Turin in Savoy on the Occasion of the Destruction of Catanea and Many Other Cities and Villages in Sicily in the Year 1692." *Philosophical Transactions of the Royal Society of London* 60 [1750]: 1–20)

the attention of some of the best scientific minds of the day. Our tour of individual historic earthquakes will begin in 1755, with the event that arguably launched not only seismology as a modern field of scientific inquiry but also the modern era of state-supported earthquake response and recovery efforts. In the annals of earthquake history, this was truly a watershed event.

By beginning the story in 1755, we of course skip over the overwhelming majority of planetary and human history, and in so doing miss some of the captivating stories of earthquakes throughout antiquity. Indeed, since the days of early man, earthquakes have shaped cultures—in surprisingly beneficial ways. Earthquakes in present-day Algeria lifted part of the Sahara, forming lakes that attracted game—verdant oases that provided food, water, and respite for early man in the midst of an otherwise hostile desert. The Dead Sea fault system similarly tilted blocks of the earth's crust, causing water to flow from mountains into the valley below, creating lakes that beckoned some 2 million years ago to mankind's earliest ancestors.

Moving from geologic to historically ancient times, one finds remnants of so-called mound cities, sites where settlements were built up, destroyed by catastrophic earthquakes, and rebuilt. Our understanding of these ancient cities, and the tales that they tell about ancient earthquakes, invariably remains murky. Both seismologists and archaeologists tend to respond with skepticism to interpretations of earthquake damage based on archaeological evidence. In chapter 2, however, we describe a few of the more compelling stories to have emerged in recent years from the synergistic field known as geoarchaeology.

As intriguing as these tales can be, this book focuses on earthquakes during the last few centuries: the earthquakes that ushered in both earthquake science and earthquake response as we know them today. It is no coincidence that the inception of modern seismology arrived arm in arm with the beginnings of the modern era of earthquake response. Neither could have developed without a certain degree of sophistication in the basic societal infrastructure. Without communal resources and the machinery to manage them, we can neither direct a recovery effort nor support the sometimes impractical, and therefore luxurious, undertakings of science.

While cogent ruminations about earthquakes date back at least to Aristotle's time, and observational seismology has mid-18th-century roots, the field of modern seismology really began to gain traction only in the mid-19th century, when Robert Mallet first discussed earthquake waves in the context of classical mechanics. When physicists speak of mechanics, they generally mean the nuts and bolts of how things behave when acted upon by basic forces such as gravity. Drawing parallels between earthquake waves and sound waves, Mallet advanced our understanding of the effects of temblors more than of the temblors themselves. Through the latter half of the 19th century a number of researchers both questioned and expanded on Mallet's seminal work, perhaps none more than John Milne, whose work was carried out at Imperial University in Tokyo. Milne's contributions to the nascent field were wide ranging, including not only theories of earthquake waves but also the nature of earthquake ruptures and the field of seismometry, the design of instruments to record earthquake waves. Primitive earthquake sensors, ranging from elegant Chinese devices to simple systems of pendulums and springs, had been invented (and reinvented) far earlier; however, Filippo Cecchi is generally credited with building the first true seismograph—an instrument capable of recording the waves from earthquakes—in 1875. At about the same time, other scientists were experimenting with similar designs, in some cases looking to record not earthquakes but tides in the solid earth; some of the early designs of these instruments proved superior to Cecchi's system, and were soon adopted for seismometers.

Not surprisingly, all of the earliest seismometer designs had significant limitations. In the annals of the development of this particular technology, John Milne emerged to play the role of Henry Ford. Having joined the Imperial College of Engineering in Tokyo in 1876, Milne joined early pioneers Thomas Gray and J. Alfred Ewing in further development of seismograph design. In Charles Davison's words, "Probably no other seismologist has had so wide an experience as he on the seismograph in all its forms."[4] Benjamin F. Howell, author of *An Introduction to Seismological Research*, called Milne "the man who was most influential in developing a practical seismograph."[5] By the late 1880s, the era of modern instrumental seismology was in its infancy, and earthquake science found itself poised to blossom by leaps and bounds.

The chapters of this book chronicle earthquakes from the 1755 temblor through the following two and a half centuries, during which both earthquake science and earthquake response came of age. Our earthquake tour includes three large events in the United States, the last of which struck northern California—most notably San Francisco—at 5:18 on the morning of April 18, 1906.

Prior to 1900, two of the most important earthquakes in the United States were, perhaps surprisingly, nowhere near California: a series of strong shocks that struck the "boot heel" region of Missouri in 1811–1812 and the Charleston, South Carolina, earthquake of September 1, 1886. While the respective scientific responses to these 19th-century seismic bookends reflect nearly a century's improvement in understanding of earthquakes, both share an important distinction. For earth scientists seeking to understand important large earthquakes, no distinction is as important as whether the earthquake is "historic," or whether it occurred during the so-called instrumental era. No matter how limited or flawed, seismometer recordings of large earthquakes confer a measure of confidence to earthquake studies; an element of ground truth to dispel the sometimes significant uncertainties that can plague historic earthquake studies. Although not a great earthquake on a global scale, the Charleston earthquake is of critical importance for understanding earthquake hazard in eastern North America—as well as similar regions worldwide—and thus ranks among the last important historic earthquakes with which scientists now contend.

If there is a downside to modern seismology, it might be that, at least for scientists who suffer from terminal curiosity, both the analysis of modern earthquake data and the results can threaten to become mundane. When a large earthquake strikes, we know how to analyze the data and, occasional exceptions notwithstanding, we will generally not be terribly surprised by the results. These days, the devil—and the new science—is often in the details. In earlier days, when scientists' understanding was more limited, each new earthquake had far more to offer by way of fundamental discovery.

Even today, scientists often have much to learn from historic earth-

Intensity	Effects
I	Not felt, recorded only by seismometers
II	Noticed by a few people
III	Many indoors feel movement
IV	Felt by many, windows rattle, noticed by some outdoors
V	Felt by nearly all, small objects overturned, plaster may crack
VI	Felt by all, slight damage to poorly built buildings
VII	Considerable damage to poorly built buildings
VIII	Damage to well-built buildings, tree branches break
IX	Considerable damage to well-built buildings
X	Most buildings destroyed
XI+	Near-total destruction, objects thrown into air

TABLE 1.1. Simplified version of the modified Mercalli intensity scale, still used by seismologists to assign a numerical value to the severity of shaking at any given location.

quakes, especially those in areas where large earthquakes are not frequent and where modern data are therefore limited. Every important historic earthquake represents a unique challenge for the scientists who endeavor to understand it. Depending on the age of the temblor, scientists can draw on more or less information about the effects of an important historic earthquake. To investigate historic earthquakes, modern scientists turn their attention to the effects of the earthquakes on people and structures, and sometimes on the earth itself.

Developed in 1902 by the Italian seismologist and vulcanologist Giuseppi Mercalli, the *Mercalli intensity scale* uses Roman numerals I–XII to quantify severity of shaking based on the effects described by witnesses (table 1.1). Such an approach is so natural that it had been reinvented any number of times prior to 1900 by scientifically inclined individuals who experienced earthquake sequences and by armchair seismologists even before the field of earthquake science existed. Armed with such a scale, scientists can now assess the distribution of damage for any earthquake that was documented by written eyewitness accounts.

At the low end of the scale, small intensities (I–II) correspond to ground motions at the ragged edge of human perception. High values are reserved for the strongest shaking possible: ground motions violent enough to substantially damage even modern, well-engineered structures. Intensity values are specific to their locations. Unlike the modern parameter, magnitude, no single intensity value characterizes the size of an earthquake. Rather, a dis-

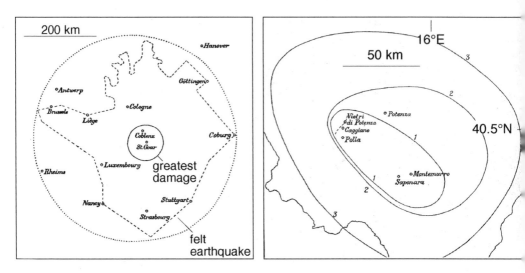

FIGURE 1.3. (Left) In 1847, Johann Nöggerath drew lines of equal shaking intensity around the epicenter of the July 26, 1846, Coblenz earthquake. (Right) Robert Mallet constructed this map to indicate his observations of damage near the December 16, 1857, Neapolitan earthquake in southern Italy. His innermost contour corresponds to Intensity X, a region where most buildings were severely damaged. (J. J. Nöggerath, *Das erdbeben vom 29 Juli 1846 in Rheingebiet und den Benachbarten Ländern* [Bonn, 1847]; Robert Mallet, *The Great Neapolitan Earthquake of 1857: The First Principles of Observational Seismology, 1862*. 2 vols. London: Chapman & Hall, 1862)

tribution of intensities can be mapped over the extent of the area impacted by an event. One of the earliest such maps was prepared by Robert Mallet to depict the shaking distribution from the 1857 Naples earthquake (figure 1.3). That intensity and magnitude are "similar but different" can lead to confusion between the two, with people sometimes concluding from an intensity map that the magnitude of an earthquake depends on location. The magnitude of an earthquake can depend on the particular magnitude scale used (there are several), but not on the location of an observer.

When scientists interpret historic accounts of earthquakes to determine intensity values, the goal is usually twofold. First, the distribution of intensity will identify the location of an earthquake, and in some cases tell us something about the extent of the fault rupture. The strongest shaking from the 1906 earthquake, for example, outlined the extent of the rupture on the San Andreas fault. Second, by developing mathematical relationships between shaking effects and earthquake magnitude for recent earthquakes, scientists produce calibrations with which the magnitudes of historic earthquakes can be estimated.

To analyze historic earthquakes, it is not enough to be a seismologist; one must become (or at least become friends with) a historian as well. The primary data for historic earthquake research are written historic accounts. These must be ferreted out of libraries or dusty archives, in some cases translated, and interpreted with an appreciation of their historic context. If someone wrote that an earthquake in 1755 knocked down his chimney and four other chimneys in town, one cannot interpret the account with any degree of confidence unless one knows something about the construction of the houses and chimneys—and whether the five damaged chimneys in town were five out of a total of seven, or five out of a total of 500.

Sometimes, careful historic research *erases* earthquakes from the catalogs. For many years the 1737 Calcutta earthquake was touted as the most lethal earthquake of all time, having apparently claimed 300,000 lives. A close review of historical facts, however, revealed it to be not an earthquake at all, but rather a cyclone-induced flood. The most telling indication of the untrustworthiness of the historic rumor was, however, the fact that the total population of Calcutta was a mere 3,000 in 1737. (To the consternation of the scientists who discover such truths, these "fake quakes" can be stubbornly difficult to erase from the record, since less careful scientists continue to propagate errors that appear in earlier references.)

When one give talks to schoolchildren it is interesting to begin with three questions: How many of you like history? How many of you like science? And, finally, how many of you like science but think that history is really pretty boring? Oddly, the third question tends to elicit more raised hands than the second, perhaps suggesting that schoolchildren consider science not wildly exciting in its own right, but at least less dull than history. Or perhaps it suggests that schoolchildren remain diffident until one says something funny. In any case, the study of earthquakes is increasingly recognized as a so-called systems science; one that requires multifaceted, multidisciplinary investigations. To study faults and modern earthquakes requires not only geology and seismology but a long list of other disciplines as well: geodesy, computer science, statistics, physics, chemistry, oceanography, and more. Sometimes history is a critical part of the mix as well. Because earthquakes happen over such long time scales, it will take a few millennia before our record of instrumentally recorded earthquakes can match the length of the current historic record. If we want to find faults, study earthquakes, and quantify earthquake hazard, the science of earthquakes must include the history of earthquakes.

Inevitably, invariably, even the most venerable human history is woefully short where earthquakes are concerned. Compared with the length scales of plate tectonics, human beings are mere ants, crawling around the surface of the planet and trying to understand the nature of features very much larger than ourselves. Compared with the time scales of plate tectonics, human beings are more like fruit flies. Damaging earthquakes recur on some plate boundaries at intervals of hundreds of years, a span of time equaling at least several human lifetimes. Nick Ambraseys has quantified the half-life of earthquake memory as 1.5 human generations in areas such as Iran. That is, children will be aware of damaging earthquakes that affected their parents, but by the next generation memories already begin to fade. In many parts of the world, earthquakes do not recur for thousands, or even tens of thousands, of years. For such areas even the invention of written language some 5,000 years ago is insufficient to provide a cultural memory of damaging earthquakes.

The grandeur of plate tectonics processes—the creation of mountain ranges and the carving of continents—defies human imagination. Even the extent of recorded history provides only the most meager snapshot in time, but our view is so limited that we cannot afford to ignore any part of it.

Accordingly, modern scientists develop methods of increasing sophistication to unlock the secrets of important historic earthquakes. In some cases, direct geologic investigations are useful as well. For intensity studies, scientists have more or less data to work with, depending primarily on the antiquity of an earthquake. The effects of the 1811–1812 New Madrid sequence were most thoroughly chronicled by a handful of scientifically inclined individuals who became impromptu naturalists, and whose findings made their way into print, and therefore to posterity, by hook or by crook. By 1886, scientists in Europe (notably Mallet, working on earthquakes in Italy) had established the tradition of the more formal and comprehensive scientific earthquake report. Early American geologists and seismologists, some of them part of the nascent U.S. Geological Survey, drew on this tradition to prepare a comprehensive and invaluable report on the 1886 Charleston earthquake.

Still, for any earthquake for which no instrumental data are available— as well as for some early earthquakes on which only limited data were collected—modern scientists must resort to ingenious seismosleuthing methods to unravel the lessons of these events. At the time important historic earthquakes struck, scientists of the day were left to sort out imperfect data from a foundation of grossly imperfect understanding. As Mallet began to describe earthquake waves, neither he nor anyone else had any real understanding of what an earthquake *was*—which is to say, the nature of the disturbance that caused the waves. And even as later scientists, including G. K. Gilbert in the late 1800s and H. F. Reid in the early 1900s, began to understand the nature of the disturbance, neither they nor anyone else had any real understanding of the fundamental forces that cause earthquakes to happen. Today, of course, we understand much more. Although we by no means understand everything, we have the luxury of looking back with an arsenal of knowledge well beyond that available to earlier generations of scientists.

One of the challenges associated with a book such as this one is figuring out how much background material to include. This is not a textbook written for students who are assumed to be at a certain well-defined point in their studies, having attended a certain set of prerequisite classes. This book is written for a range of readers, including those who have some background in earth sciences and a curiosity about historic earthquakes, as well as those who know little about earthquakes in general. It is impossible to craft some parts of this book for the latter group without boring the socks off of the former, but at this juncture we can offer a bit of advice for readers. The next section presents a brief overview of plate tectonics and basic "earthquake ABCs," and can very safely be skipped by readers who have read about these topics before.

Faults and Plates

Before focusing on earthquakes and faults, one must deal first with the planet on a continental scale. As most people know, the occurrence of earthquakes is now understood within a paradigm known as plate tectonics. Some of the ideas associated with this theory, such as continental drift, are centuries old, but an integrated and mature theory did not appear on the scene until the middle of the 20th century. According to the basic tenets of plate tectonics theory, the upper layer, or crust, of the earth is broken into about a dozen major plates. These plates intersect at one of three boundary types: spreading, subduction, or transform (lateral) zones. The crustal plates ride atop the mantle, in which the ongoing process of convection is accommodated by gradual flow. Earthquakes cannot occur in the mantle because rocks are too hot and plastic to develop faults. An imperfect but conceptually reasonable analogue for the mantle is sold in brightly colored plastic eggs: Silly Putty. (Contrary to popular perception, truly molten, or liquid, rock is scarce in the earth; it is found only in the outer shell of the earth's core and in small pockets in the crust.)

The upper layers of the crust, on the other hand, are known to be brittle. Like the best-known brittle gemstone, diamond, the crust is breakable under the right circumstances but is otherwise very strong. (Among the myriad ways that the crust does not resemble diamonds: the former does not break along geometric cleavage planes.) The formidable strength of the crust allows the enormous plates to remain intact. Motion within the crust occurs at the boundaries of plates, not, by and large, within them. At subduction and transform boundaries especially, plates generally move past each other in abrupt lurches we know as earthquakes. Earthquakes and faults come in three basic flavors: lateral, or strike-slip; thrust, and normal. Although simple cartoon depictions of plate boundaries suggest that motion is concentrated along one primary plate boundary fault, earth scientists generally speak of plate boundary zones. That is, while one principal fault—such as the San Andreas fault in California—might account for most of the plate boundary motion, secondary faults in the region inevitably exist and account for an appreciable fraction of the long-term motion. In California, the San Andreas fault thus does not simply slide one side of the state past the other; rather, a system of faults is, effectively, cutting a broad swath of the state into ribbons. Earthquakes such as the 1992 M7.3 Landers and the 2002 M6.6 San Simeon, California, temblors therefore represent motion along the plate boundary, even though they occur well away from the San Andreas fault. Complex plate boundary zones are the norm elsewhere in the world as well—for example, Iran, where the devastating 2002 Bam earthquake struck away, although not too far away, from the country's most active earthquake zones (figures 1.4 and 1.5).

In any active plate boundary region an additional set of secondary faults inevitably accommodates secondary stresses associated with the primary faults. Again, a classic example can be found in California, where a bend in the San Andreas fault generates a broad compressional, or squeezing, force across much of southern California (figure 1.6). This compression drives a complex assemblage of faults—both thrust and strike-slip—that produce earthquakes such as the M6.7 Northridge temblor of 1994. In a broad sense an active plate boundary zone can therefore fairly be considered earthquake

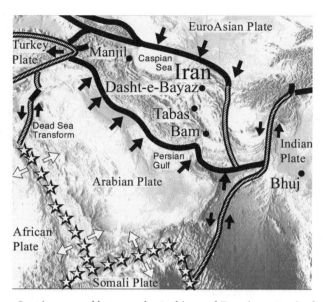

FIGURE 1.4. Iran is squeezed between the Arabian and Eurasian tectonic plates. Stars indicate spreading plate boundaries, bold lines represent convergent boundaries, and striped lines show transform faults. Arrows indicate the sense of relative motion. The locations of recent earthquakes that each killed more than 20,000 people are named on the map. The 2002 Bam earthquake occurred in southeastern Iran, well away from plate boundaries.

country, because earthquake hazard will affect a much larger area than the narrow strip immediately adjacent to the primary plate boundary fault. Overall, active plate boundaries account for only a small fraction of the total real estate on the planet. It might seem like more, however, because of the aforementioned tendency of plate boundary regions also to be some of the most densely populated areas on earth.

Active plate boundaries are found along the full extent of the western coast of North America: a subduction zone offshore of Alaska, strike-slip faults offshore of northern British Columbia, another subduction zone offshore of the Pacific Northwest, the strike-slip San Andreas fault system through California, a mixed zone of strike-slip and extensional faulting through northern Mexico, and yet another subduction zone offshore of most of Mexico. The so-called Pacific Rim Ring of Fire comprises these zones and more, each of which generates large earthquakes at a healthy (or, rather, unhealthy) clip. There are other active plate boundary zones in populated regions elsewhere around the globe. From the Himalaya collision zone in northern India, an active plate

FIGURE 1.5. A typical domed roof near the town of Tabas in Iran that is being assembled from mud bricks and mud mortar only two weeks after the M7.5 1978 earthquake that destroyed most of the town's similar adobe structures. Each year, an outer coating of mud will be added to increase thermal insulation, resulting in a roof that is thicker than 50 centimeters and that weighs many tons. More than 200,000 people have died in Iran since 1890 from the earthquake-induced collapse of these traditional adobe houses. (From Roger Bilham, *Seismological Research Letters* [2004]: 706–712)

boundary zone continues westward through Iran, Turkey, Greece, Italy, and eventually to Lisbon and the mid-Atlantic. A long subduction zone runs along the entire coast of Chile, and a continuous belt of active plate boundaries, including subduction zones, surrounds the Caribbean. The 2004 Sumatra quake occurred on a subduction zone along the eastern edge of the India plate.

Not every coastline is an active plate boundary. The Atlantic coast of North America, for example, separates the North American continental crust from the Atlantic oceanic crust, but this boundary is what geologists term passive. North America is drifting away from Africa and Europe because new oceanic crust is being generated along the Mid-Atlantic Ridge. The North American continental crust gets pushed sideways along with the oceanic crust to the east of the coast; there is no relative motion between the two, and therefore relatively few earthquakes. Thus eastern North America is quite staid, geologically speaking, compared with the dynamic environment on the western side of the continent. The overall rate of earthquakes on the two sides of the continent differs by as much as a factor of 100.

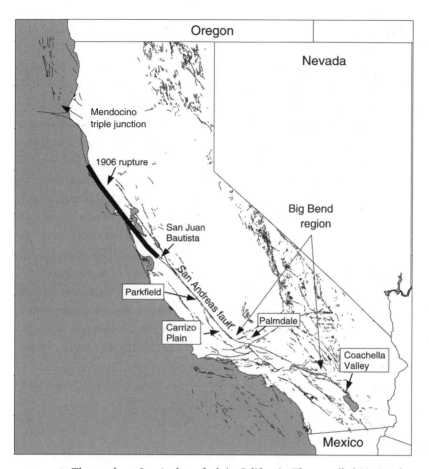

FIGURE 1.6. The southern San Andreas fault in California. The so-called Big Bend causes compression across the greater Los Angeles region, causing the transverse ranges to grow—a growth process that occurs primarily in the paroxysms known as earthquakes.

Yet quiet does not mean dead. Both the historic record and several lines of evidence reveal that potentially large earthquakes do occur along the Atlantic seaboard. The most notable such event in historic times was the 1886 Charleston, South Carolina, temblor, discussed in chapter 6. A more moderate but still damaging earthquake offshore of Cape Ann, Massachusetts, in 1755 caused widespread chimney damage in the young colony (see sidebar 1.1). As later chapters will discuss, we do not fully understand the forces that cause these sorts of earthquakes, nor can we say with confidence where they will, and will not, occur in the future. But even a passive continental margin is

SIDEBAR 1.1

The great Lisbon earthquake of 1755, discussed in chapter 3, struck on the morning of November 1. The Cape Ann earthquake, thought to have been the largest historic temblor in the state of Massachusetts, struck 17 days later, damaging chimneys, gable ends, and stone fences in Boston and Cape Ann. Prior to 1992, most scientists would have said with conviction that it was only a fluke that these two temblors struck in such close succession. Following the 1992 Landers earthquake in southern California, however, seismologists recognized a new class of earthquakes: remotely triggered earthquakes. These events are like aftershocks in that they are caused by a large main shock, but they occur at much greater distances. Looking back at historic earthquakes, it is sometimes possible to find compelling evidence for remotely triggered earthquakes even without data from seismometers. Sometimes, as in the case of Lisbon/Cape Ann, we don't have enough information to prove a link between the temblors, but we certainly can no longer say with conviction that no link exists. From an earthquake science point of view, the planet emerges as a more dynamic and interesting—although also perhaps sometimes more dangerous—place than scientists recognized just a few years ago.

a relatively complicated place by virtue of the transition from thick continental crust to thin oceanic crust. Where there is marked variability in the structure of the crust, there are invariably stresses of some sort. The occurrence of earthquakes in such a region should not surprise us.

At this juncture it is perhaps useful to review briefly the relationship between earthquakes and faults as understood in modern times. The geologic definition of "fault" originated in mining, where the term was applied to manifest fractures, or flaws, in otherwise unbroken rock. The association between geologic faults and earthquakes was not made until the very end of the 19th century. Prior to this time, evidence of faulting had been observed in conjunction with earthquakes, but the upheaval of real estate was generally considered to be the result rather than the cause of earthquakes. The confusion is understandable in retrospect, since the upheaval caused by earthquakes can include far more than the actual motion on a fault. Large earthquakes can cause large tracts of land to slump, typically toward bodies of water; they can also trigger massive landslides. Observable fault ruptures can pale in comparison with the so-called secondary effects of earthquakes.

On October 28, 1891, a large earthquake struck Japan, claiming over 7,000 lives and leaving a conspicuous scar, or surface rupture, that could be traced about 100 kilometers across the landscape. Professor Bunjiro Koto of Imperial University described the feature and concluded that the great rent was the actual cause of the Mino-Owari earthquake. This was arguably the first such deduction, which now represents one of the most fundamental tenets of earthquake science: an earthquake is an abrupt motion along a surface known as a fault. (As is so often the case in science, few ideas are entirely without precedent: in 1819 famed geologist Charles Lyell observed that low-lying land had risen during the 1819 Allah Bund earthquake in western India.) The great 1906 San Francisco earthquake, which left a conspicuous surface rupture some 470 kilometers long, further cemented scientists' understanding of the nature of earthquakes—as discussed in chapter 7.

People often think of earthquakes as being located in a particular place—a not unreasonable inference, given our tradition of naming earthquakes after particular cities or other geographical features. Yet the discussion above highlights an important point: earthquakes do not happen at points, but along faults. For the most part, the magnitude of an earthquake depends on the extent of fault area that moves during the event. The M9.0 2004 Sumatra quake ruptured a mind-boggling 1,600 kilometers of fault (roughly the length of California plus Oregon); the M8.1 1999 Tibet earthquake ruptured 400 kilometers of fault; the 2002 M7.9 Denali Park, Alaska, earthquake extended about 300 kilometers along the Denali fault; and the 1992 M7.3 Landers, California, earthquake left a jagged rupture in the desert some 70 kilometers long. In modern times, investigations of earthquakes and faults are inextricably intertwined. How this state of understanding evolved is an important part of the overall development of modern earthquake science.

Returning to the question of where earthquakes occur, from a geologic point of view most midcontinent regions, including the heartland of North America, are even more sedate than the passive Atlantic margin. From the eastern front of the Rocky Mountains to the eastern continental margin, there is the vast and largely unbroken extent of the North American plate. As anyone who has ever driven coast to coast knows, it's a big country out there. The cornfields alone go on forever. And for the most part the country is geologically of a piece. The North American plate, like the earth's other major tectonic plates, does not experience substantial internal deformation, but remains an intact puzzle piece which moves only relative to other puzzle pieces. Much of the midcontinent is flat; mountains such as the Appalachians owe their existence to plate tectonic forces that were active many millions of years ago. In their current intraplate environment, such mountains are being worn down by gradual but inexorable erosional forces, with no active tectonic forces pushing them up. Yet big earthquakes can occur in the heart of the heartland as well. The 2001 Bhuj earthquake occurred several hundred kilometers from India's active plate boundaries, in a region where large earthquakes are expected to be relatively infrequent. The understanding of such midcontinent, *intraplate* earthquakes, and the faults on which they occur, has lagged well behind the understanding of earthquakes and faults along active plate boundaries.

The complexity of intraplate earthquakes and faults poses a challenge to scientists, and remains a lively and intriguing issue in the earth sciences. And when considering the societal implications of earthquakes in relatively inactive parts of the world, it is important to remember the distinction between *hazard*, which reflects a region's exposure to earthquakes, and *risk*, which reflects a region's exposure to damage given the hazard. That is, the earthquake hazard along the San Andreas fault is without question high, but if one pitched a canvas tent and set up camp immediately adjacent to the fault, one's risk would be quite low. Californians do not generally live in tents; however, they tend to live in structures designed and built to withstand earthquake shaking. In other parts of the world, where earthquakes have been infrequent during our short historic record, the existing inventory of buildings contributes substantially to risk, even if new structures are built to stringent codes.

As Nick Ambraseys first observed many years ago, earthquakes don't kill people, buildings kill people. The issue of building vulnerability poses enormous challenges in many countries around the world, especially developing nations. Yet a sizable percentage of structures in places such as the eastern United States and the United Kingdom, including many in areas where large earthquakes have occurred during the short historic record, are not designed to withstand earthquakes. As scientists endeavor to understand earthquakes and faults in parts of the world that are not known as earthquake country, the hazard implications are never far from our minds.

The dual theses of this book may seem oddly juxtaposed: societal response on the one hand, the development of scientific thinking on the other.

Yet this marriage of themes feels entirely natural to earthquake scientists: we spend our careers pursuing interests whose practical implications are as important as their academic implications are intriguing. Steven Jay Gould observed that "curiosity impels, and makes us human."[6] When it comes to natural phenomena that have societal consequences, curiosity alone defines neither humans nor scientists; nor is it all that impels us. But as inherently curious creatures, it is impossible for humans not to be fascinated by the amazing planet we call home.

And so, keen to understand earthquakes as both a natural phenomenon and a natural hazard, we find ourselves with marching orders as daunting as they are clear: scientists must investigate and try to understand earthquakes with the barest snapshot of time and the most meager of data. This book is, in part, about scientists' efforts to rise to that challenge, efforts that involve a unique marriage of concern for societal issues and often ingenious, fascinating science. It is also about the remarkable, ongoing journey of scientists to understand the planet that is our home, from the standpoint of a very young species that has come of age in a very old world.

2

Earthquakes and Ancient Cities: Armageddon—Not the End of the World

It was built against the will of the immortal gods, and so it did not last for long.
  —Homer, *The Iliad*

The reduction of an entire city to a pile of rubble poses a special problem for the survivors. Roads are blocked, underground pipes are broken, and disease accompanies the decay of incompletely buried bodies. Fresh water and sewage no longer flow, food becomes scarce, and the absence of shelter from extremes of temperature can make life miserable.

In the cities of the ancient world a very real practical problem followed in the months and years after the destruction of a city—a cleanup operation beyond the wildest dreams of the survivors. Although steam shovels had been used for moving heavy materials in building the Suez and Panama canals in 1869 and 1910, respectively, it was not until 1923 that the bulldozer was invented (figure 2.1). The even more useful backhoe followed 25 years later.

Thus, clearing debris was a daunting task as recently as the 1906 San Fran-

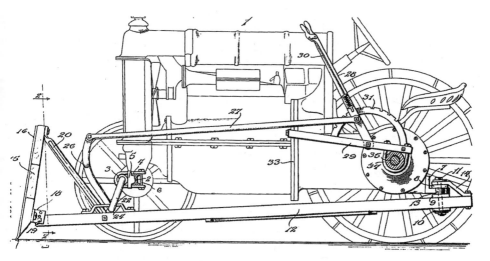

FIGURE 2.1. A sketch of the first true bulldozer, invented in 1923. Prior to this time, removal of large quantities of earthquake debris required a Herculean effort that was sometimes beyond the logistical capabilities of early civilizations. (From J. E. McLeod and J. D. Cummings, United States Patent and Trademark Office Publication Number 01522378. Filed December 18, 1923; issued January 6, 1925)

cisco earthquake. In his book *The City That Is: The Story of the Rebuilding of San Francisco in Three Years*, Rufus Steele wrote of the rebuilding effort:

> First the ground had to be cleared. The task would have baffled Hercules—cleaning out the Augean stables was the trick of a child compared to clearing for the new city. This is a step in the rebuilding which fails entirely to impress the visitor of today. He can form no conception of the waste which had to be reduced to bits and then lifted and carted away to the dumping grounds. The cost of removing it was more than twenty million dollars.[1]

Lacking what we would now consider modern machinery to move large volumes of debris, the rebuilders of San Francisco extended railway lines across town, brought in steam and electric cranes, and relied heavily on teams of horses that suddenly found themselves in enormous demand. According to Steele, "Huge mechanical devices for shoveling and loading were invented and set to work."

Formidable as the task may have been, San Francisco tapped into several critical resources in its Herculean efforts: trains, cranes, and, perhaps most important, large numbers of survivors following an earthquake that killed a very small fraction of the local population. The situation was far bleaker after any number of devastating earthquakes in earlier times. The cleanup following the ruinous destruction of cities prior to 1800 was undertaken strictly by hand. Where there were insufficient survivors of the earthquake, it could not be undertaken at all, and the old city would remain a ruin, the rubble slowly becoming a shapeless mass.

Only a few cities have been left in this dismal state, and they are all in the ancient world. In modern times, the abandoning of a collapsed but otherwise viable city is extremely rare. In ancient times, collapsed cities had considerably more inertia. Whereas in a few days a bulldozer can pack a collapsed high-rise into giant trucks to be tossed into the Sea of Marmara (as happened following the Izmit earthquake of 1999), or to be piled into artificial hills (as happened following the ChiChi, Taiwan, earthquake of 1999), the ancients had no bulldozers, and the collapse of a city would have posed a cleanup problem exceeding the capabilities of survivors to solve. A hundred square kilometers of masonry or mud blocks, combined with the incineration of wooden structures in an earthquake-induced fire, would have been beyond both imagination and experience, even in times of war. Adding proverbial insult to injury, in the immediate aftermath of a catastrophic earthquake in ancient times, not only may a city be destroyed but also, sometimes, the veneer of law and order that stabilizes day-to-day life was disrupted. Looting sometimes broke out covertly or violently, discouraging families from reestablishing their homes.

Sufficiently severe damage to a city mandates its temporary abandonment, with survivors moving to the outskirts or staying with relatives in distant towns, as has occurred after recent earthquakes. The outbreak of disease due to the collapse of water supplies ensures the isolation of the city for many months. But like a magnet the natural amenities of a city draw survivors back to recreate something of its former glory. The geographical advantages that drew people to a location in the first place are rarely altered in any significant way by an earthquake; inevitably these same advantages beckon people to return. And home is home, even when it's a mess.

Following large earthquakes in recent historic times, there has been a process of resettlement almost as predictable as the biological process known as old-field succession, whereby plant life returns to a region following a devastating forest fire. After 20th-century earthquakes in Iran, for example, the slow return of survivors to their former homes began in the least damaged areas.

Where the mess was too difficult to manage, villagers erected new structures on the outskirts of the town, or on hurriedly leveled ground, in the weeks and months following the disaster. Religious buildings and schools were repaired for community use. Improved structures followed on repairable existing foundations, often built with materials cannibalized from destroyed buildings. Lean-tos and temporary roofs were replaced as prosperity increased, at a rate moderated by the availability of water or by the influx of urban engineers with the vision and administrative support to reconstruct roads, civic buildings, and drainage systems. Slowly but inexorably cities returned to life.

A local disaster can be fixed with regional help. But what if the earthquake is sufficiently severe to wipe out a nation's infrastructure? A disaster of such proportions might be highly unlikely in modern times, but nations were more vulnerable in ancient times. A direct hit near an administrative capital could destroy not only the buildings but also societal mechanisms to regulate the peace. With nowhere for survivors to go, and no hope of rescue or support from a larger entity, recovery could be slowed and even halted. With food no longer flowing along former supply channels, city dwellers unaccustomed to foraging would be forced to leave for distant parts where their skills could be put to use. In the absence of gainful employment, the former workers of a society would need to revert to life-supporting activities: scribes to hoeing, musicians to digging, businessmen to laboring.

Substantial regional disruption was possible in the ancient world when an earthquake or a series of earthquakes disrupted the food chain, and removed the leadership and infrastructure needed to impose orderly recovery. For example, the earthquake that damaged the Galilean port of Tiberias on January 18, 749, took with it 60 neighboring cities over a region measuring 90 kilometers by 170 kilometers. The largest empires of the ancient world created the largest cities of the ancient world. If these largest cities were severely damaged, the empire could be dealt a lethal blow.

Written records of ancient earthquakes tend to be scarce, but sometimes a temblor's effects are preserved within the ruins themselves. In Bet She'an, a city in the Jordan Valley with a rich history dating back to biblical times, an earthquake in A.D. 749 is recorded by linear arrays of collapsed columns, side-tilted arches, and dislodged and pulverized masonry. Under one of the columns, archaeologists discovered one skeleton reaching for coins. Excavations of this abandoned city revealed the occasional cannibalized column, apparently recycled after an earthquake had severely damaged the city in A.D. 361. After the 361 earthquake the city recovered. In 749 it died.

The fault that slipped in 749 may also have damaged the city of Megiddo, otherwise known as Armageddon. A fault runs right through Armageddon: geophysicist Amos Nur has proposed (although not without controversy) that multiple earthquakes on this fault were responsible for the biblical connotations of the word. Earthquakes there in 1500 B.C., 1000 B.C., and 500 B.C. may have destroyed this garrison town, leading to our modern metaphor for the end of the world. The world, however, will not end with an earthquake: as later chapters discuss, every earthquake is part of an elastic rebound process that will ultimately lead to the next earthquake. Future earthquakes will continue to damage the site, as well as other locations along the plate boundary between Africa and Arabia.

Adobe Domes, Tells, and Ancient Mounds

Around 9,000 years ago the 10,000 Neolithic citizens of Çatalhöyük, near the modern city of Konya, in Turkey, constructed their homes with bricks made from mud. And to this day adobe construction is still used when no other materials are available and where the climate is predominantly dry. Bricks cut from wet clay and stacked to dry in the sun have the advantage of being free; they can be glued together with more wet mud, providing a finished wall of great uniformity, if dubious toughness. Toughness (the resistance of a material to fracture) is one of the most desirable of earthquake-resistant attributes, and the toughness of adobe can be improved by mixing the mud with straw, a practice known to the Egyptians. Although strength can be greatly improved by baking the mud, the resulting bricks are easily fractured with a few deft blows of a hammer. Moreover, a pile of bricks is no stronger than the mortar that glues them together.

Many of the world's poor cleave adobe bricks from unadulterated wet clay. Adobe walls can be finished to a plasterlike flatness and covered with protective paint on the outside and with comforting paintings on the inside. Once four walls have been erected, they have sufficient mass and strength to support a domed roof. A skilled worker can build an arch of adobe blocks with no intermediate supports, starting at the corner formed by two walls and progressing to the opposing corner. The finished structure insulates against extremes of temperature and the occasional downpour. Blemishes can be repaired with more wet mud, and each year a new layer of mud smeared on the outside provides additional thermal insulation.

An adobe arch roof typically starts with a membrane of four-inch-thick bricks. A smooth layer is placed over its outer surface after it has dried, and, if the owners are wise, after each rain shower or drought. Without attention, unfired mud brick structures revert to nature and their edges become blurred after a few years. Rain smoothes the details, and aridity carries crumbling surface layers away as dust.

After a few generations of loving maintenance, the thickness of a roof will have increased to two feet or more, and the roof can weigh many tons. In earthquake country the weight of such structures literally becomes their downfall. During a powerful earthquake the walls are unable to resist the massive horizontal forces caused by side-to-side shaking, and once weakened, the entire structure collapses to become its owner's tomb. If the adobe roof breaks into large chunks, half a dozen rescuers are needed to shift the pieces with their bare hands. (Crowbars, poles, and ropes are of course buried and unavailable.) If the rescue involves just one collapsed dwelling, crushed survivors can be released, but when a whole city collapses, there is insufficient time for rescuers to release victims from the many thousands of tons of collapsed debris. Rescue operations become a frantic race against time as survivors search for signs of life. The scene now plays out on television following damaging earthquakes: seemingly against all odds, people are pulled alive from the rubble days after a temblor strikes.

Tragically, however, even dramatic rescue is no guarantee of survival. We now know that immediate deaths from an earthquake often constitute a small fraction of the final death toll. Many more die in the following days due to crushed muscles, as the liver tries to remove the products of muscle damage from the body. The death toll in the hour following an earthquake can increase tenfold in the following five days unless the bruised survivors are fed salt-rich solutions.

We know all too well what the aftermath of a modern earthquake looks like. But how do we recognize it in archaeological sites? At Bet She'an the evidence is obvious: systematically toppled columns, crushed bones. However, the evidence is not always so clear. When scholars consider an ancient city that reveals signs of devastation, the discovery of human remains beneath fallen ruins is regarded as the hallmark of an ancient earthquake. Where the collapsed structures consist of massive stonework, there is little doubt about what crushed the skeletons. But the case is not so clear in the mounds, or tells, of ancient adobe cities. The Çatalhöyük ruins contain numerous burial layers within the dwellings of its mound, but the skeletons are not crushed and the burials are interpreted as an orderly social or religious procedure. However, the absence of crushed bones does not necessarily mean a person did not die in an earthquake.

Another telltale sign of those few seconds of shaking that can destroy a city is the presence of unpillaged valuables—not just gold or silver, or family heirlooms, but food and tools that an invading army would surely have removed before destroying a captured city.

But by far the most important calling card of an earthquake is the offset of the earth's surface by the faults that moved during the quake. Such evidence is rare because it requires the unintentional construction of dwellings or fortifications directly atop the fault responsible for the earthquake. On May 20, 1202, however, a Crusader's castle was demolished just north of Tiberias by slip in an earthquake. The castle walls were offset 1.6 meters by this earthquake and by a further 0.5 meter by earthquakes in October 1789. These later earthquakes were responsible for the collapse of three of the nine remaining pillars of the Roman temple of Jupiter at Baalbeck observed in 1751. Dozens of columns had collapsed in previous earthquakes.

In most cases, though, archaeological ruins tell an ambiguous story at best. We will probably never know just how many cities of the ancient world were razed by earthquakes. If one earthquake has damaged a city, however, it is reasonable to assume that other quakes may have been responsible for previous or succeeding layers of a city's chronology. Although archaeologists will protest that not every layer of a mound city can be attributed to an earthquake, seismologists can respond with the observation that earthquakes are far more common than wars. It is reasonable to conclude that many more archaeological ruins owe their survival to the ravages of earthquakes than archaeologists currently believe. In the same way that Pompeii was preserved by volcanic ash, earthquakes have provided time capsules for present-day archaeologists.

Many of the multiple layers of Jericho likely owe their origin to repeated shaking because the city lies near active faults that even now are visited by earthquakes. With its lowest levels dated at 9000 B.C., Jericho is believed to be the longest continuously inhabited city in the world. The Jericho mound is about 15 meters high and contains more than 20 levels of destruction and construction. The biblical account of the walls of Jericho tumbling down might reflect one of the several earthquakes that must have damaged or destroyed the city since its founding.

The city of Taxila, near present-day Islamabad in Pakistan, lies in the foothills of the Himalaya, a region where great earthquakes are caused by the northward motion of India beneath the Tibetan plateau. Three shifts in the city's location have occurred in the past 3,000 years, and although no specific earthquakes have been identified as causing its collapse, we know that earthquakes exceeding M8 must occur one or more times each millennium.

Kathmandu, Nepal, was damaged by an M8.1 earthquake in 1934, and the rubble from its reconstruction was recently exhumed during the installation of sewers beneath the present city. Under the debris of 1934 lies the rubble of earlier destructive earthquakes, in 1255 and in 1833.

No mound city is as famous, or as romantic, as the one built and rebuilt on a fortuitously located limestone ridge along the Aegean Sea just below the Dardanelles: Troy. Nine layers of destruction have been recorded in Troy, which prior to excavation was a mound of dirt some 100 meters above the surrounding fields. Excavation showed the first five levels (Troy I–Troy V) belong to the early Bronze Age (3000–1900 B.C.). Troy VI was built on a larger scale but destroyed by an earthquake sometime after 1300 B.C. Troy VIIa was hurriedly constructed from the ruins of Troy VI, and appears to have been destroyed by an invading army around 1240 B.C., possibly the army of Agamemnon. Reconstruction (Troy VIIb) was followed by apparent abandonment for some 400 years until Greek settlement led to a new phase of construction (Troy VIII) around 700 B.C. The Greek settlement was destroyed by the Romans in 85 B.C. but reconstructed (Troy IX), only to be abandoned after the founding of Constantinople about A.D. 300.

The Troy of myth and legend—Homer's Troy—was Troy VI, of the middle-to-late Bronze Age. The Trojan War lasted for ten years before the sack of Troy. Is it possible that a modest earthquake assisted the city's end—an accident of nature exploited by an opportunistic army? It wouldn't be the first time that an earthquake has changed the course of history. An earthquake in the night of September 1, 1803, presumably weakened the resolve of the well-defended Maratha fort of Alygarh in India, which was surrounded by a British army preparing for a siege lasting several months. The British, in tents outside the city defenses, were unaffected by earthquake shaking (from an M8.2 earthquake 150 kilometers to the north, in the Himalaya), whereas the city dwellers were distracted, confused, and eventually overwhelmed by the British attack. The story of the Trojan horse may indeed have been a metaphor for Poseidon, the Greek god of earthquakes. Could the attacking Greek army have taken advantage of earthquake damage to the wall of Troy, or of the confusion or distress of its inhabitants?

Amos Nur points to evidence that a series of strong earthquakes contributed to the demise of Bronze Age civilizations around 1200 B.C. In his words, an "earthquake storm" struck the eastern Mediterranean and Near East between approximately 1225 and 1175 B.C.: "the greatest catastrophe in what would become western civilization."[2] Along with Troy, the ancient cities of Mycenae and Knossos might have met their end during these turbulent decades. While Nur's theories remain highly controversial, substantial evidence suggests that the fate of mankind's ancient cities was sometimes shaped by earthquakes in fundamental ways.

If the immortal gods gave any thought to earthquake risk, Troy certainly would have been built against their will. Its seemingly advantageous location notwithstanding, western Turkey is one of the planet's most prolific earthquake zones. The Northern Anatolian fault, which cuts across northern Turkey and was responsible for the devastating Izmit earthquake of 1999, is thought to branch into several roughly parallel faults through the Sea of Marmara region and points westward. These fault branches pass north, south, and essentially straight through the location of ancient Troy (figure 2.2).

The vast majority of ancient mound cities lie in earthquake belts in the Middle East, along the Silk Road, and in the collision belts of Iran and Afghanistan, India and China. Yet some tells and mounds lie many hundreds of kilo-

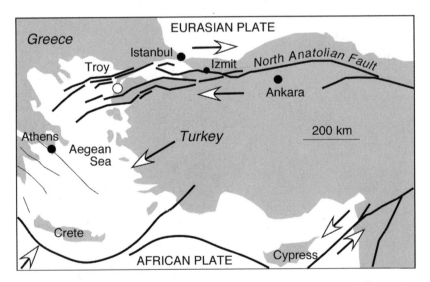

FIGURE 2.2. The North Anatolian fault system runs through northern Turkey and beneath the Sea of Marmara. Splays of the fault continue westward, putting the ancient city of Troy in the crosshairs.

meters from regions where we think earthquakes are common. What of the mound cities of central India, or places such as Jenne Jeno, south of Timbuktu in the central Saharan kingdom of Chad, or the mounds of North America?

Whereas many of these may have been abandoned because of geographic circumstances, some of them may be linked to the occasional midplate earthquake. In the central United States, the Mississippian culture flourished for centuries in southeastern Missouri, southern Illinois and Indiana, western Kentucky and Tennessee, and parts of northeast Arkansas. Chief among the hallmarks of the Mississippian tradition was mound building: the deliberate creation of large, mostly earthen mounds—ceremonial centers in elaborate cities whose populations numbered well into the thousands (figure 2.3). Ceremonial earthen mounds clearly were not created by earthquakes, but some evidence suggests that they were created *because of* earthquakes. A high concentration of mounds is found in the Cairo Lowland, the corner of Missouri tucked against Kentucky to the east and Illinois to the north. In this area, the 1811–1812 New Madrid earthquakes are known to have caused substantial liquefaction: in particular, dramatic features known as sand blows, which represent a literal venting—even fountaining—of sand to the surface during strong shaking.

Two lines of evidence suggest a link between earthquakes and mounds. First, as originally noted by geologist Roger Saucier, some excavations reveal sand blows and fissures at the bases of mounds, suggesting that the mounds were deliberately built atop features created in earlier earthquake sequences. Geologists have compelling evidence that the New Madrid region was struck by sequences of two to three large earthquakes (and their aftershocks) in 900 and around 1450. It seems almost self-evident that early civilizations would have interpreted prolonged earthquake sequences, and the spewing of sand and other material from the bowels of the earth, in spiritual terms.

A second line of evidence suggesting a link between earthquakes and early cultures is the striking coincidence of dates: archaeologists trace the start of the Mississippian culture to 900, precisely when the region was rocked by one of the earlier sequences of earthquakes. Geologist Martitia Tuttle has spent years investigating and determining ages of New Madrid liquefaction features, the latter process sometimes relying on archaeological artifacts to date geologic strata. She has been struck over the years by the remarkably close correspondence between seismic upheaval and cultural upheaval. There is another intriguing correspondence as well: archaeologists

FIGURE 2.3. A large earthen mound of the Mississippian culture in Cahokia, Illinois. (Courtesy of Cahokia Mounds Historic Site)

have shown that sometime near the beginning of the 15th century, many of the temples in the Cairo Lowland were burned—whether on purpose or accidentally, they cannot say. Probably we will never be able to say for certain, but if earthquakes were responsible for the razing of Mississippian temples, it would be neither the first nor the last time that cultural sites of spiritual significance met their ends in this way.

Remembering What Not to Do

The tells and mounds of the ancient world tell a story of urban settlements: locations where in some cases cities were almost certainly destroyed by earthquakes, and where modern cities are sitting ducks for the damaging temblors yet to come. In what we now call the earthquake belts of the world, many of them the cradles of civilizations, there are hundreds of mound cities. Why, one may ask, did these early city dwellers never learn that top-heavy structures fare poorly in earthquakes, that masonry and mud bricks collapse into rubble, and that certain locations are more vulnerable to damage than others?

The simple answer is that earthquakes happen infrequently enough that there is no possible way for most builders to form general conclusions about an appropriate response. In fact, most city dwellers do not suppose that an earthquake is one of a series of recurring quakes that stretches back into pre-history and, more important, stretches with an inevitable certainty into the future. Few builders construct buildings thinking that *their* construction will in fact be the death of its occupants in some future earthquake, even if one has just occurred.

Earthquake-resistant construction requires an intellectual appraisal of structural shortcomings, often based on a library of knowledge extending over many generations and geographical settings. Before the 20th century this information was unavailable to most architects and contractors. Even now it requires special laws to alert the unwary builder to the importance of earthquake resistance. Ancient peoples, with virtually no awareness of earthquake science or earthquake engineering, can scarcely have interpreted a damaging temblor in anything but spiritual terms: the "act of God" that could be neither predicted nor prevented.

With earthquakes as with everything else, the farther back in time one attempts to gaze, the murkier the details. Earth scientists struggle to establish the most basic parameters for the earliest known earthquakes: when, where, and how big? But if the earthquakes themselves seem vexingly inscrutable, their impacts on people and societies clearly are hazier still. As the story of Troy exemplifies, ancient history can blend almost seamlessly with lore and legend. Archaeology is a science different from geology or seismology: the historians of the distant past take on the daunting challenge of evaluating scant physical evidence and making sense of human behavior and motivations. Such interpretations are by their very nature speculative.

Still, curiosity impels us to look back in order to understand where we came from. The cradle of mankind may be in Africa, but when early hominids took their first steps away from the nursery, they walked along a corridor created by plate tectonics: the Dead Sea fault zone, along which the motion of two plates carved a lush and inviting valley in the midst of an otherwise arid terrain. Emerging from this valley, early man established the cradle of Western civilization in the heart of one of the planet's fiercest earthquake zones. The continued northward motion of Africa, Arabia, and India created active earthquake belts stretching from China to the Mediterranean.

Iran, Turkey, Greece, Italy—the great centers of early Western civilization— were rocked by severe earthquakes again and again—and in some regions, by volcanoes as well. The limitations of archaeological evidence and inference notwithstanding, we know beyond reasonable doubt that the course of early civilizations was influenced—sometimes even destroyed—by natural catastrophes: the Minoans by the devastating eruption of Santorini, the Romans by the eruption of Vesuvius. Not yet so clear is the cumulative influence of swarms of earthquakes along the Middle Eastern earthquake zones that may have interrupted the smooth development of Bronze Age cultures.

Combining this handful of compelling cases with the absolute certainty that large earthquakes will recur regularly in active plate boundary zones, the earth scientist cannot help but suspect that earthquakes influenced ancient cultures more strongly than archaeologists have recognized. A new paradigm has emerged: *natural disasters shape cultures.* There is a tendency for this assertion to be expressed as it is here, in the present tense—that is, to draw implicit parallels between what we know, or think we know, about the ancient world and what is true in modern times.

As every physical scientist knows, however, extrapolation is often fraught with peril. The earth scientist knows that earthquake rates in different regions around the globe have remained the same, in a long-term, average sense, since *Homo sapiens* first arrived on the scene. The anthropologist knows that *Homo sapiens* has remained fundamentally the same species throughout its period of existence. But one part of the equation has changed considerably since ancient times: the nature of societies and civilizations. The advent of modern civilization, with its attendant mushrooming of population, clearly has left humans more resilient in some respects but more vulnerable in others.

How, then, have earthquakes and other natural disasters impacted societies in the recent historic past? In the chapters that follow, we explore this question, the answer to which must be gleaned from careful consideration of large earthquakes that have struck during historic times. To truly understand the impact of these earthquakes, one must understand the earthquakes themselves—and, in turn, the development of understanding in earthquake science. Thus we bid farewell to the ancient world and, in the next chapter, set the clock forward to the middle of the 18th century: the year 1750. It was at this time that small rumblings presaged far greater tumult throughout Europe: for seismology as well as civilization, truly a watershed moment.

3

The Lisbon Earthquake
and the Age of Reason

Sapere aude!
— Immanuel Kant

An individual's response to any catastrophic event, including the capacity for rebound, surely depends a great deal on one's expectations before the event. In a short-term sense, earthquakes remain as utterly unpredictable and abrupt as they have been since the dawn of time. Looking back through history, however, it becomes apparent that some earthquakes were more unexpected—and seemingly more mercurial—than others.

In the middle of the 18th century, earthquake science had barely reached its infancy. Earthquakes had fascinated, and posed a challenge to, the best minds since at least the day of Aristotle. Aristotle, Pliny the Elder, St. Thomas Aquinas—whether they viewed earthquakes as acts of God or not, they and other philosophers approached the subject with a decidedly naturalist bent. Aristotle and Pliny interpreted earthquakes as the result of subterranean winds or subterranean storms. St. Thomas Aquinas argued in favor of the scholastic approach, supporting Aristotle's scientific views over later, more theologically oriented interpretations.

During the 17th century, earthquakes continued to be the source of scientific speculation. Galileo argued that the earth had a dense, solid core. In 1680 Robert Hooke published *Discourse on Earthquakes*, arguably the first significant book dealing with earthquakes as a natural phenomenon.

In 1750 a series of earthquakes was widely felt throughout England. During this "year of earthquakes,"[1] shocks were felt in London on February 19 and March 19, in Portsmouth and the Isle of Wight on March 29, in northwest England and northeast Wales on April 18, and in and around Northamptonshire on October 11. These shocks are now estimated to have been no larger than mid-magnitude-4: the first two events were quite small, felt strongly in London only because their epicenters were within city limits. But pound for pound—or, rather, magnitude unit for magnitude unit—the impact of these earthquakes far outstripped their literal reverberations within the earth. As Charles Davison recounts, nearly 50 earthquake-related articles were communicated to the Royal Society by the end of 1750. (The publication process must have been quite a bit speedier in 1750 than it is today.) Lists of British and global earthquakes were published in the *London Magazine* and the *Gentleman's Magazine*. The appendix of one publication, thought to have been written by the Rev. Zachary Gray, included accounts of 61 destructive earthquakes worldwide and descriptions of 41 earthquakes felt in England between 974 and 1750.

Notwithstanding these and other efforts, prior to the time of the Lisbon earthquake in 1755, what was really known about earthquakes could essentially be summed up in two words: practically nothing. The temblor that rocked Europe in 1755 could scarcely have been less anticipated; even given scientists' current understanding of earthquakes, the Iberian Peninsula scarcely stands out among the great earthquake zones of the world, or as a likely candidate to produce a great earthquake. Indeed, even having produced a great earthquake, the source region of the Lisbon earthquake remains enigmatic. In a broad-brush sense, the European and African plates are converging toward one another, creating a wide, diffuse zone of faulting that includes subduction (thrust) faults. A preponderance of evidence— including the dramatic tsunami—points to an offshore, thrust-fault source. Scientists (Eulalia Gracia and others) have identified the 100-kilometer-long Marques de Pombal fault (MPF) in this region. This fault is too small, however, to produce a temblor as large as the magnitude estimated for Lisbon, a whopping 8.5. To appreciate the magnitude of the magnitude, it is helpful to consider the logarithmic nature of the magnitude scale: the fact that an M8 is about 30 times larger than an M7. At M8.5, the Lisbon temblor was a substantially larger beast than the 1906 San Francisco earthquake, which had an estimated magnitude of 7.9.

But we are putting the cart before the horse. Returning to the morning of November 1, 1755—a Sunday—witnesses describe a day shining bright and glorious: "never a finer morning seen," according to the Rev. Charles Davy, at the beginning of one of the most careful and complete first-person accounts of the events that followed.[2] Between 9 and 10 o'clock in the morning, Davy felt a gentle motion that grew to a trembling of the house, first "imputed to the rattling of several coaches in the main street."[3] When these rattlings were followed by a "strange frightful kind of noise under ground," Davy grew concerned that the phenomenon might presage an earthquake—from his account it appears that he did not consider the initial rattlings and rumblings to actually *be* an earthquake.

Realizing that the initial disturbances might be the forerunner of something worse, Davy contemplated whether it would be safer to remain in his apartment or run to the street. His opportunity for considered thought ended abruptly when he was "stunned with a most horrid crash, as if every edifice in the city had tumbled down at once. The house I was in shook with such violence, that the upper stories immediately fell; and though my apartment (which was on the first floor) did not then share the same fate, yet everything was thrown out of its place in such a manner that it was with no small difficulty I kept my feet, and expected nothing less than to soon be crushed to death."[4] Making his way to the street through choking clouds of dust, Davy encountered a scene of utter chaos. Climbing over the ruins of St. Paul's Church, he found "a prodigious concourse of people of both sexes, and of all ranks and conditions, among whom I observed some of the principal canons of the patriarchal church . . . several priests who had run from the altars in their sacerdotal vestments in the midst of their celebrating Mass; ladies half dressed, and some without shoes; all these, whom their mutual dangers had here assembled as to a place of safety, were on their knees at prayers"[5] (figure 3.1).

In the midst of these feverish devotions, a "second great shock came on, little less violent than the first, and completed the ruin of those buildings which had already much shattered." This second shock was, moreover, "attended with some circumstances still more dreadful than the former. On a sudden I heard a general outcry, 'The sea is coming in, we shall all be lost.'" Turning toward the Tagus River, Davy observed the waters "heaving and swelling in the most unaccountable manner," and then, "In an instant there

FIGURE 3.1. Sketch of damage to St. Paul's Church following the 1755 Lisbon earthquake. (From p. 522 of A. H. Godbey, *Great Disasters and Horrors in the World's History*. St. Louis, Mo.: Imperial Publishing, 1890)

appeared, at some small distance, a large body of water, rising as it were like a mountain. It came on foaming and roaring, and rushed towards the shore with such impetuosity, that we all immediately ran for our lives as fast as possible; many were actually swept away, and the rest above their waist in water at a good distance from the banks"[6] (figure 3.2).

Returning to the river, Davy observed ships tossed about as if in the midst of a violent storm, and even greater horrors: "The fine new quay, built entirely of rough marble, at an immense expense, was entirely swallowed up, with all the people on it, who had fled thither for safety, and had reason to think themselves out of danger in such a place: at the same time, a great number of boats and small vessels, anchored near it (all likewise full of people, who had retired thither for the same purpose), were all swallowed up, as in a whirlpool, and nevermore appeared."[7] Witnesses described the river level rising 6 meters and then abruptly falling, causing the quay and vessels to be swallowed up whole, seemingly into a cavity. One witness concluded that the

FIGURE 3.2. Artist's rendition of tsunami and devastation caused by the 1755 Lisbon temblor. (From p. 520 of A. H. Godbey, *Great Disasters and Horrors in the World's History*. St. Louis, Mo.: Imperial Publishing, 1890)

cavity must have instantly closed, for "not the least sign of a wreck was ever seen afterwards."

Lisbon would be rocked by a third large shock that morning, somewhat less severe than the first two, but also accompanied by the dramatic sea effects that scientists now recognize as a tsunami.

Davy contemplated his predicament: clearly it was dangerous to remain at the shore, yet the houses inland offered little in the way of safe haven. He decided to go to the Mint, a "low and very strong building" that "had received no considerable damage." There he found the soldier guards gone save for one officer, no more than 17 or 18 years old, who informed Davy that "Though he were sure the earth would open and swallow him up, he scorned to think of flying from his post." Davy credits this man's bravery and sense of duty with preventing the robbery of the Mint, "which at this time had upwards of two millions of money in it."[8]

Although the third strong shock and second tsunami marked the end of the remarkable initial sequence, the inevitable secondary disaster—fire— soon followed. All Saints' Day was a high festival in Portugal, and "every altar in every church and chapel . . . was illuminated with a number of wax tapers and lamps as customary." Toppled candles ignited toppled timbers and curtains, and "the conflagration soon spread to the neighboring houses, . . . being there joined with the fires in the kitchen chimneys." As darkness fell, "the whole city appeared in a blaze, which was so bright that I could easily see to read by it. It may be said without exaggeration, it was on fire at least in a hundred different places at once, and thus continued burning for six days together, without intermission."[9]

As is often the case, estimates of the final death toll varied, but Charles Davy reckoned it to be more than 60,000 in Lisbon alone—over 20 percent of the city's population of 275,000. The Royal Palace was destroyed by the earthquake and subsequent tsunami. Royal archives documenting early explorers, paintings by the likes of Titian and Rubens, a 70,000-volume library—centuries of accomplishment were wiped out in a few minutes of terror.

Moreover, the devastating effects of the Lisbon earthquake stretched far beyond the city. The temblor caused severe shaking in North Africa, with damage and loss of life in the cities of Fez and Mequinez, and moderate damage in Algiers, over 1,100 kilometers distant. The tsunami caused damage along the coasts of Portugal, southwest Spain, and western Morocco. In the Algarve region of southern Portugal, witnesses described waves reaching 30 meters high. Although the tsunami were responsible for much of the damage away from the Lisbon area, shaking from the temblor was strong enough to be experienced in France, Switzerland, and northern Italy. Vesuvius had been erupting prior to the earthquake; when the earth shook, the volcano stopped. Scientists now know that large earthquakes can affect the plumbing systems of large volcanoes, but in most modern cases for which we have good observations, the impact is in the other direction: earthquakes disrupt a volcanic system and trigger activity, not cause it to cease.

When the Lisbon earthquake literally rocked Europe, it metaphorically rocked the lofty philosophical debates of the 18th century. The Age of Enlightenment had begun around the start of this century, with "enlightenment" defined by Immanuel Kant as "Man's release from his self-incurred tutelage. Tutelage is man's inability to make use of his understanding without direction from another."[10] Kant further declared the motto of the age to be *Sapere aude!*, which translates literally as "Dare to be wise!" and less literally as "Have courage to use your own reason!" Thus defined, enlightenment was not inherently inconsistent with religion, but the tenets of this concept certainly led philosophers to challenge dogma and to apply their most critical logic skills to issues that had previously (generally) been matters of faith alone. Echoing reasoning proposed by the classical philosopher Epicurus (c. 341–270 B.C.), Voltaire famously summarized one aspect of the debate as follows: "God can either take away evil from the world and will not; or being willing to do so, cannot; or He neither can nor will; or, lastly, He is both able and willing. If He is willing to remove evil and cannot, then He is not omnipotent. If He can, but will not remove it, then is He not benevolent; if He is neither able nor willing, then is He neither powerful nor benevolent; lastly, if both able and willing to annihilate evil, how does it exist?"[11]

For Voltaire, the devastating and utterly unheralded Lisbon earthquake represented a prime case in point. As argued by Colin Brown in *Christianity and Western Thought*, the earthquake reinforced Voltaire's belief that "God is not affected by evil. Neither does he concern himself with human wretchedness and misery. There is no Garden of Eden or Paradise to be restored."[12]

As a larger movement, however, enlightenment was more than an attack on established religion. Although not a wholehearted devotee of enlightenment, Jean-Jacques Rousseau to some extent reconciled many of the age's tenets with his acceptance of a Judeo-Christian God, as did Leibniz and Pope, for whom earthquakes represented imperfections in a natural world created by God. Kant viewed God as beyond the purview of logic and reason, concluding that the reality of God could be neither proved *nor disproved* "by merely speculative reason."[13]

One cannot be surprised by such disparate views, considering again the motto that Kant proposed for the age: in its essence, enlightenment was about questions, not answers. When considered from this viewpoint, the Lisbon earthquake had an effect far beyond reinforcing the views of any individual philosopher *or* religious leader. (Following the temblor, John Wesley, a pioneer of the 18th-century Evangelical movement in England, was among those who argued forcefully that earthquakes resulted from the "hand of the Almighty" rather than "purely natural and accidental" causes.[14]) Even as philosophers and religious leaders struggled to understand the earthquake in the context of their views, scientifically inclined individuals—some of them religious leaders as well—took small but important steps toward a systematic, scientific exploration of earthquakes as a natural phenomenon. If the earlier "year of earthquakes" in Great Britain had primed the pump, so to speak, kindling scientific interest in earthquakes, the Lisbon temblor let loose the floodwaters. Some of the most scientifically astute minds of the day found themselves striving to document and understand earthquakes. In doing so, they essentially affirmed some of the ideas they sought to test: simply by embarking on a systematic exploration of a natural phenomenon, they were implicitly recognizing it to *be* a natural phenomenon, one that could be understood through observation and rational deduction.

Together with the 1750 shocks, the Lisbon earthquake thus became the catalyst for seismology as a field of scientific inquiry. Although earthquake catalogs had been compiled prior to 1755, in particular in 1750, cataloging efforts were redoubled after the event. In Switzerland, naturalist Elie Bertrand (1712–1790) compiled a chronological account of earthquakes felt in Switzerland from 563 to 1754. This and other early cataloging efforts got the ball rolling, so to speak, inspiring the efforts of others in the years that followed. In England, Robert Mallet compiled a catalog of known earthquakes between 1606 B.C. and A.D. 1842, a Herculean effort that was published between 1852 and 1854 in a series of reports nearly 600 pages long.

As discussed in chapter 4, Mallet's seminal contributions played an important role in the establishment of seismology as a modern science. But one should not underplay the importance of the contributions made by Mallet's predecessors. Bertrand and John Michell went well beyond a simple chronicling of known earthquakes. Along with their contemporaries, these men began to endeavor to interpret observations in a systematic, scientific manner. Bertrand was among the first to recognize the high speed of earthquake shaking; he also recognized that earthquakes struck more commonly in some parts of Switzerland than in others, a difference he attributed to the presence of mineral springs, caves, and sulfur beds in earthquake-prone areas.

In England, Michell considered the distribution of earthquakes on a broader scale. Considering the record of earthquakes over long and short timescales, he noted a difference between earthquakes that recur over large intervals and the closely spaced earthquakes that follow a large temblor — perhaps the first formal recognition of main shocks and aftershocks. Referring to Chile and Peru, he observed that regions near "burning mountains" experienced more earthquakes than other areas. And, presaging a recognition that would be made by the modern seismological community only in 1992, he noted that the 1755 Lisbon earthquake was succeeded by smaller local shocks in Switzerland and elsewhere.

Michell went on to consider the nature and cause of earthquakes. Inevitably his theories missed the mark, appealing in large part to "subterraneous fires." "These fires," he wrote, "if a large quantity of water should be let out upon them suddenly, may produce a vapour whose quantity and elastic force may be fully sufficient for the purpose."[15] Michell also described the wavelike motion of earthquakes as follows: "Suppose a large cloth or carpet (spread upon a floor) to be raised at one edge and then suddenly brought down again to the floor, the air under it, being by this means propelled, will pass along until it escapes at the opposite edge, raising the cloth in a wave . . . as it goes."[16] Although Michell viewed this mechanism as a means to move large quantities of "vapor," in many respects these words presaged a slip-pulse model for earthquake rupture that was proposed by seismologist Thomas Heaton in 1990.

Together with the 1750 shocks, the Lisbon earthquake heralded the age of observational, empirical seismology. These earthquakes captured the attention of naturalists such as Michell, Bertrand, Mallet, and their contemporaries— some of the best scientific minds of their day. This phenomenon—the galvanization of intellect by dramatic natural events—is ubiquitous throughout the history of science. Invariably, the most lively minds will be captured by the most interesting, and the most immediately pressing, problems of the day.

Philosophers and religious leaders contemplated the Lisbon earthquake in the context of their philosophical theories. Apart from the issues that defined the Age of Enlightenment, the 18th-century Western world was also one in which *reason* played an increasingly important role. By this time, Sir Isaac Newton had established some of the most fundamental tenets of physics, Copernicus and Galileo had challenged the geocentric view of the universe, and Kepler had derived elegantly simple rules to predict the motions of the planets. A predictable and orderly world began to emerge: the behavior of all objects, from falling stones to giant planets, could be understood, and *predicted*, from immutable scientific laws.

If human beings had begun to view the universe as orderly and predictable, one can scarcely imagine a bigger philosophical earthquake than the temblor that devastated Lisbon and rocked much of Europe. A devastating earthquake striking out of the blue on All Saints' Day, toppling candles from church altars to ignite a citywide conflagration—one struggles to imagine an earthquake that would appear to be more obviously an act of God. Yet, as illustrated by Voltaire and John Wesley, individuals considered the same event and reached diametrically opposed conclusions: on the one hand, that God does not exist, and on the other hand, that humans cannot hope to understand fully the ways of God. Yet even as the earthquake challenged theories and beliefs, it became a critically important catalyst for the observational, empirical, and *reasoned* consideration of the natural world. (Many point out that *reason* was scarcely a new invention in the 18th century; this point notwithstanding, it is inarguably also clear that man's views of the natural world have evolved gradually from more religious and more superstitious times.)

Like other cities, Lisbon did rebound from the devastation wrought by the earthquake and subsequent fire. As argued by Russell Dynes, "the major consequence of the Lisbon earthquake was perhaps not in the intellectual debate it generated but, for the first time, an embryonic modern state assumed collective responsibility for disaster consequences. This was a precursor of later actions by modern states to bear some of the consequences of disaster losses. While the Enlightenment argument had lasting consequences, the acceptance of state responsibility was a more lasting social invention."[17]

The Marques de Pombal, Portugal's prime minister, directed the rebuilding of the city, favoring a simple architectural design now known as Pombaline. The wide, parallel avenues in the Baixa district were rebuilt after the earthquake, and are one of the city's attractions to this day. The earthquake did exact a steep toll from the city, however. Prior to 1755, Lisbon was the fourth largest city in Europe, a port famous for its commerce and wealth. Gold strikes in Brazil helped make the first half of the 18th century a prosperous era in Lisbon. Although the city rose from the ashes following the earthquake, it would never regain its former power and prestige on the world stage. (Napoleon's four-year occupation in the early 19th century certainly did nothing to help, nor did the fact that Portuguese preeminence in maritime commerce had already begun to wane by the time the earthquake struck.) Today, Lisbon is a vibrant city and an important center of commerce, but it is dwarfed in size not only by London and Paris, but also by cities such as Vienna, Prague, and Kiev.

As later chapters will discuss, the rebound of regions following major earthquakes is largely due to a focusing effect. In particular, major temblors focus resources, not only from individual countries but also, often, from all over the world. The Lisbon earthquake helped usher in the modern era of state-coordinated and state-supported disaster recovery, but, as is often the case with trailblazers, the city of Lisbon did not benefit from the trail to the same extent as cities that followed in its wake.

The city of Lisbon did recover, however, and the wide-ranging direct impact of the earthquake was matched by the remarkable and wide-ranging indirect impact of the event. The earthquake did little to settle the great philosophical debates of the 18th century: Does God exist? If He exists, why do bad things happen? Is the natural world—of which earthquakes are a part—predictable? One suspects that at least some of these debates will never be settled. But even as it challenged the Age of Enlightenment and Reason, the Lisbon earthquake inspired some of the best thinkers of the day to use observations to consider and understand the earthquake as a natural phenomenon. Perhaps more than any other single seismic event, the temblor was responsible for the development of seismology as a modern field of scientific inquiry. And in a larger sense it provided a substantial impetus for the movement, which was already well under way, to consider the natural world in scientific terms rather than in a context of religion or superstition.

When scientists describe the earth's crust as elastic, we mean that it has the ability to bend without breaking and then, essentially, snap back into place. Elastic rebound is, however, just one of the consequences of elasticity. Seismologists who study earthquake ruptures speak of overshoot, by which they mean the fact that a fault rupture sometimes goes beyond the stops, just as a bent sapling will swing beyond its point of equilibrium before returning to its usual upright stance. Elasticity also implies a range of consequences in a human or societal context. People and cities rebound from earthquakes, but given the elastic and dynamic nature of social systems, dramatic events such as catastrophic temblors can also provide a measure of momentum—a challenge, a sense of energy; sometimes an impetus for change.

Modern estimates of the magnitude of the Lisbon earthquake peg the event as well below the size of the truly great earthquakes that have rocked the planet in historic times, including the 1960 quake in Chile, the 1964 Good Friday temblor in Alaska, and the 2004 Sumatra event. But perhaps no other temblor generated more bang for the buck, if bang is considered in terms of overall impact. The Lisbon earthquake played a major role in launching seismology as a modern field of scientific inquiry; it also helped change the way that human beings perceive their relationship with the natural world. Charles Davy and others who experienced the Lisbon earthquake described a horrific ten minutes of shaking in the initial shock. Ten minutes can be an eternity when the ground is shaking fiercely enough to topple buildings; ten minutes can be nothing at all when considered as a unit of time. And once in a great while, ten minutes can change the world.

4

Tecumseh's Legacy:
The Enduring Enigma
of the New Madrid Earthquakes

There is nothing, I believe, so trying to a healthy nervous system
as a succession of earthquakes.
> —Louis Housel, "An Earthquake Experience,"
> *Scribner's Monthly* 15, no. 5 (1878)

Prologue: Tecumseh's Prophecy

Like any proper mystery, the tale of the New Madrid earthquakes begins on
a note of intrigue. According to legend, the earthquakes were predicted—
even prophesied—by the great Shawnee leader and statesman Tecumseh
(figure 4.1). Concerned over continued encroachment of white settlers onto
Indian lands in the midcontinent, Tecumseh traveled widely throughout the
central United States in the early 1800s, striving to unite diverse tribes to
stand against further land cessions.

According to legend, Tecumseh told his mostly Creek followers at Tuckabatchee, Alabama, that he had proof of the Great Spirit's wrath. The sign
blazed across the heavens for all to see—the great comet of 1811, a dazzling

FIGURE 4.1. One of the few reportedly contemporary portraits of Shawnee leader Tecumseh, based on a pencil sketch made by Pierre Le Dru during the War of 1812. (From B. J. Lossings, *Pictorial Field-Book of the War of 1812.* New York: Harper and Brothers, 1868)

and mysterious sight. As if to emphasize Tecumseh's words, the comet grew in brilliance through October, dimming in the nighttime sky in November just as Tecumseh left Tuckabatchee for points northward.

Also according to legend, Tecumseh's speech at Tuckabatchee told of an even more dramatic sign yet to come. In an oration delivered to hundreds of listeners, the leader reportedly told the crowd, "You do not believe the Great Spirit has sent me. You shall know. I leave Tuckabatchee directly, and shall go straight to Detroit. When I arrive there, I will stamp on the ground with my foot and shake down every house in Tuckabatchee."[1] The Creeks counted the days until the one calculated to mark Tecumseh's return, and on that day— December 16, 1811—the first of the great New Madrid earthquakes struck, destroying all of the houses in Tuckabatchee.

Tecumseh's Prophecy, as it has come to be known, strikes a chord with those inclined to see Spirit and earth as intertwined. But it can also capture the imagination of those who see phenomena such as earthquakes as the exclusive purview of science. What if Tecumseh's Prophecy was born not of communication with the Great Spirit, but instead of an ability to recognize signs from the earth itself? According to the renowned English geologist Sir Charles Lyell, Native American oral traditions told of devastating earthquakes in the New Madrid region prior to 1811. What if Tecumseh had heard this legend, and more? What if Indian legends told not only of the earth-quakes themselves, but also of physical signs that had presaged the earlier events? Perhaps an observer as astute as Tecumseh had recognized those same changes during his extensive travels in the midcontinent. Perhaps, then, earthquakes are more predictable than we now believe . . . if only one is aware enough to recognize the signs.

Or perhaps Tecumseh's Prophecy is no more than a tall tale, born of imag-ination and romance rather than either science or spirit. Although Tecum-seh's speeches often were recorded by American and British observers who closely followed the leader's activities, a thoroughly researched modern biog-raphy by John Sugden concludes that no reliable record exists of Tecumseh's speeches during the fall of 1811. Tecumseh is known to have arrived in Tuck-abatchee on September 19, 1811, but, unhappy to find white observers pres-ent, refused to address the crowd for some days. When U.S. agent Benjamin Hawkins left, Tecumseh finally delivered his speech to the pan-Indian coun-cil; a speech at which no U.S official was present, and of which no written record exists.

According to Sugden, Tecumseh seems to have exhorted his listeners not to speak of the speech or their plans. Contemporary accounts of the Tucka-batchee address are vague, suggesting that Tecumseh sought to rally the Southern tribes to join his confederacy and stand united against land ces-sions. Beyond this, little is known.

However, better documentation exists for a speech Tecumseh gave to the Osages in late 1811 or early 1812; a speech that, by all accounts, was given after the December 16 temblor. As recounted by John Dunn Hunter in his mem-oirs, Tecumseh told his audience:

> Brothers, the Great Spirit is angry with our enemies. He speaks in thunder, and the earth swallows up villages, and drinks up the Missis-sippi. The great waters cover their lowlands. Their corn cannot grow, and the Great Spirit will sweep those who escape to the hills from the earth with his terrible breath.[2]

Oral storytelling may be notorious for its unreliability, but sometimes writ-ten storytelling is no better. The date of Tecumseh's speech to the Osages has been erroneously identified as the *spring* of 1811, and thus has been consid-ered by some to be Tecumseh's first prophecy. Clearly it was no such thing. Tecumseh probably did believe the Great Spirit was behind the extraordinary natural events of 1811. He likely would have been quick to draw his followers' attention to the tapestry of the natural world, a tapestry he believed had been embroidered by the hand of the Great Spirit: Look at this comet; look at this earthquake—these represent signs from the Great Spirit, proof that the Great Spirit has sent me. In the absence of plausible scientific explanations for either phenomenon, Tecumseh's words would surely have resonated deeply. Even today, the scientist can only appeal to coincidence—never a very gratifying explanation, even when it is surely right—to explain why the great comet and great earthquakes occurred so close together, and at such a critical juncture in the history of American settlement.

But what of the so-called second prophecy? The one so rich with detail? History remains mute on whether the words were indeed spoken at Tuckabatchee; with no definitive record, one cannot rule out the possibility that they were. One must remember, though, that history is less mute on the subject of what was said and done after the earthquakes. According to Sugden, by spring of 1812 it was being reported that Tecumseh had predicted the earthquakes. Far more telling, however, were the words of Tecumseh himself, as documented by John Dunn Hunter and others. The leader so passionate about his cause, so believing in the signs and their meanings—surely he would not have passed up the opportunity to claim credit . . . had he in fact made a successful earthquake prediction. But listen again to Tecumseh's documented words in the immediate aftermath of the first shock: "The Great Spirit speaks in thunder, and the Earth swallows up villages." Nowhere does one find words such as "as I foretold." If the prophecy was in fact made, Tecumseh would be perhaps the only individual since the dawn of time to have made a successful earthquake prediction . . . *and not bothered to take credit for it.*

One surviving contemporary account of the earthquakes—the "New Madrid Extract," discussed later in this chapter—includes a secondhand report from an Indian who said that "the Shawnee Prophet has caused the earthquake to destroy the whites."[3] "Shawnee Prophet" was the name given to Tecumseh's brother, Tenskwatawa, who usually traveled with the leader. Although again tantalizingly suggesting a prophecy, the statement stops short of a claim that the earthquake had been predicted. And the Indians were far from unanimous in associating the temblor with Tecumseh or his brother. Another contemporary observer, John Wiseman, recounted the explanation he had heard from a different Native American in the New Madrid region: "Great Spirit, whiskey too much. Heap drunk."[4] In the end, only one credible conclusion can be drawn from what was said and, equally important, not said: the prophecy never occurred.

One should be careful, however, of the lessons drawn from the saga of Tecumseh's Prophecy. The legend cannot detract from the man, who by all accounts was an extraordinary statesman and orator, a match for any produced by the more developed societies of that era or any other. In fact, a careful reading of history suggests that the legend—the fact that it is a legend rather than fact—should only enhance the stature of the man, precisely because he did *not* lay claim to a prediction he had not made. As modern seismologists know only too well, many lesser mortals have over the ages fallen before the same temptation. And so we are left with the story; a good yarn with just enough truth to be credible and just enough mystery to capture the imagination. The story could scarcely be anything but captivating, given its lead characters: Tecumseh and the New Madrid earthquake sequence of 1811–1812. The former was one of the most charismatic and romantic figures of American history, and the latter, one of the most important and perplexing earthquake sequences to have struck during historic times. But if Tecumseh's Prophecy is an enigma, the New Madrid earthquakes are the proverbial riddle wrapped inside an enigma. And far more than intellectual curiosity hinges on the solution to this particular riddle. The need to assess future seismic hazard in the midcontinent provides an unwavering imperative to assemble the evidence—sparse and ambiguous as it may be—and make sense of it.

By virtue of the antiquity of the evidence, scientific investigations of the New Madrid sequence have required as much ingenuity as acumen; as much seismosleuthing as seismology. As arguably the most important sequence during the short history of the United States, New Madrid has been the focus of considerable attention from the seismological community—including, in recent years, one of the authors of this essay. Thus has New Madrid become a veritable seismosleuthing poster child—that is to say, testimony to the amount of information that can, with ingenuity and acumen, be gleaned from preinstrumental observations. Because of both its importance for U.S. earthquake hazard and the fact that it is near and dear to the authors' hearts, we chronicle this journey of adventure and science in some detail.

We also address the issue of societal impact, as the New Madrid story is intriguing and richly layered in this regard as well. After all, the earthquake sequence rocked the midcontinent just a handful of years after the Louisiana Purchase; just as settlers of European descent had begun to stake their claims in the frontier region.

The legacy of the 1811–1812 New Madrid earthquakes is thus one of both riddles and romance, curiously paralleling that of the great Shawnee leader who may not have prophesied, but who certainly bore eloquent witness to, one of the most remarkable earthquake sequences in U.S. history.

The Setting

Understanding what happened at New Madrid over the long, cold, unsettled winter of 1811–1812 requires an appreciation of context, historical as well as geologic.

Historically, the New Madrid region was at something of a crossroads in 1811. Although sold to the United States by Napoleon in 1803, the Louisiana Territory had remained a Spanish possession until 1802. Passage along the Mississippi River, and especially through the port of New Orleans, had been a contentious issue between Spain and the United States throughout the 1700s. Seeking to establish a strategic line of defense in the late 1700s, Spain offered land along the west side of the Mississippi and, critically, the promise of free trade along the river, to Americans, in the hope that these settlers' allegiance would then remain with Spain.

Frustrated in attempts to secure a land grant in more settled parts of the nascent country, former patriot George Morgan was among those who seized the opportunity offered by the Spanish. Morgan set sail down the Ohio River in 1788, eventually choosing a location along the Mississippi some 100 kilometers below its confluence with the Ohio. His site was known at the time by a French name derived from the abundance of bear and buffalo in the region, L'Anse a la Graisse. The name sounds far more elegant in French than its literal English translation: "Cove of Grease." Perhaps mercifully, Morgan took it upon himself to change the name to New Madrid, in an attempt to curry favor with the king of Spain and help convince him to confirm the land grant.

Conceived on a grand scale, New Madrid enjoyed urban planning well ahead of its time. Morgan laid out the town with carefully and sensibly numbered streets and designated locations for schools, churches, the king, and the poor. There was a perhaps surprising side to George Morgan, land baron: he laid out his town with an eye toward—and an appreciation of—open spaces, wild animals, and Native Americans. Substantial tracts of land, including some along the river, were set aside for parks, and trees anywhere in the city could not be cut down without permission from a town official. Out of respect for both Native Americans and the animals, Morgan decreed that hunting would be strictly limited to killing for each family's own consumption.

In addition to its forward-looking planning and environmental conservation, New Madrid was blessed with seemingly substantial advantages associated with its location. Situated on the outside of a horseshoe bend in the river and perched atop a bank standing well above the water, New Madrid possessed a commanding view, its location exuding a sense of stability. New Madrid quickly became one of the two most important boat landings along the Mississippi, second only to Natchez farther south.

Unfortunately, the apparent advantages conferred by the natural setting proved illusory. The banks—so critical to head off the ever-present danger of floodwaters along a major river—turned out to be an ephemeral feature of the landscape. As early as 1796, French general Victor Collot observed, "The river which by its direction strikes with force upon this perpendicular bank, carries away, at different periods of the year, a considerable quantity of ground on which the town and fort are built."[5] The first settlers also found the area, with its poorly drained land, markedly unhealthy.

Some 190 kilometers north of New Madrid, residents of the French settlement of Sainte Genevieve encountered a similar lack of hospitality along the banks of the great river. Originally situated immediately along the river within *le grand champ* (the big field)—the town suffered damage from floods as early as 1780. These assaults paled in comparison to those of 1785, still known by many Sainte Genevieve residents as *L'année des grandes eaux*: the year of the great flood. Upon arriving in town after the flood, a keelboat captain set out to survey the damage and found only a stray chimney top and roof ridge visible above the water. When the waters receded, settlers returned to the mess left behind: decaying fish and livestock, foot-deep mud inside of homes. Salvaging what they could, they established a new town site inland, on *les petites côtes* (the little hills) of limestone that begin just a mile west of the river (figure 4.2). The modern limits of Sainte Genevieve stand in testimony to the peril associated with proximity to the Mississippi River, for although the town has grown westward from the historic town center, it has never grown east to reclaim the original town site (figure 4.3).

New Madrid, however, stayed put . . . and foundered. By 1797, George Morgan and many of the original settlers had thrown in the towel and returned home. Several factors may have contributed to Morgan's decision to leave. The Spanish government had imposed additional, more onerous re-

FIGURE 4.2. The first brick building west of the Mississippi, a survivor of the
1811–1812 New Madrid sequence. This handsome structure is now a restaurant
in the historic town of Ste. Genevieve. (Susan E. Hough)

strictions in the years after New Madrid was first settled. Morgan also inher-
ited a sizable estate in a more settled and hospitable part of the country. New
settlers arrived, but by 1810, Henry Marie Brackenridge observed the district
of New Madrid to be "but thinly populated, considering the great proportion
of fine land which it contains."[6] Scottish naturalist John Bradbury passed
through a year later, in December 1811, and found something less than a thriving metropolis: "I was much disappointed in this place, as I found only a few
straggling houses situated around a plain of from two to three hundred acres
in extent. There were only two stores, which are very indifferently furnished."[7]

    An examination of early settlement patterns reveals a consistent general
pattern. With land transportation difficult and costly, the earliest settlers of the
late 1700s and very early 1800s remained clustered in proximity to the waterways. Settlement expanded not in a uniformly advancing front but as a web,
with the major rivers—the Ohio, Mississippi, Missouri, and St. Francis—as

its limbs. By the 1820s, patterns had begun to change, however, with the recognition that a river's largesse often comes at a steep price: not only floods but also disease.

The year 1811 saw a frontier on the cusp of transition, with increasing awareness of the hazards associated with the early settlement patterns but the new patterns not yet established. What happened over the winter of 1811–1812 can only have further eroded settlers' confidence in the hospitality of the Mississippi River Valley. One might imagine that the winter therefore played a significant role in shaping future settlement along America's greatest river.

Or did it? Before answering this question, let us set the stage geologically. Although 19th-century settlers did not know that most large earthquakes occur at active plate boundaries, the modern reader does bring this awareness to the story. This then begs the question, Why did large earthquakes occur in such a seemingly unlikely location?

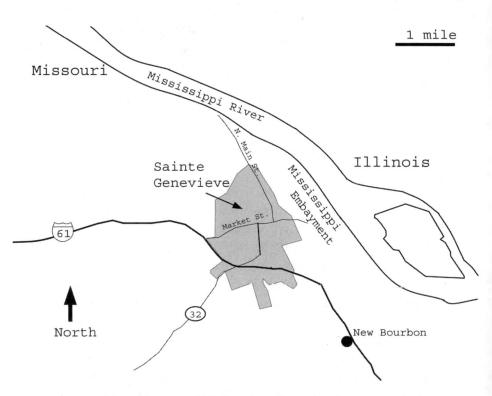

FIGURE 4.3. Map of the town of Ste. Genevieve. Even today, the town remains inland of the flood-prone Mississippi embayment, immediately west of the river.

Historically the Mississippi River Valley was at something of a crossroads in the early 19th century. In geologic terms, it has been at something of a crossroads for hundreds of millions of years. The New Madrid seismic zone represents a scar, if you will, within the old, stable continental crust of the North American midcontinent. Hundreds of millions of years ago, tectonic processes acted to pull the crust apart in this location, forming what geologists know as a *rift*. Such processes typically remain active for millions of years, but not forever. Tectonic plates can and do change directions and/or rates of motion at times; active processes can and do come to an end. In the midcontinent, the rifting process died out some 500 million years ago, leaving a zone of weakness that persists to this day—a failed rift. According to one current model, the deep crust within the New Madrid seismic zone is weak not because the rock type is inherently weak, but because the rocks are hotter, and therefore more yielding, than the surrounding colder crust. If the midcontinental crust of North America is a chain, the New Madrid seismic zone is the weakest link (figure 4.4). As is often the case in science, other scientists have proposed other explanations. We will, however, spare the reader the gory details.

The tectonic forces now acting upon the midcontinent are paltry by the standards of active plate boundary regions such as California, Alaska, and Japan. Yet forces are at work virtually everywhere on the planet, and central North America is no exception. In and around New Madrid, several different forces shape the crust. These forces include a broad "push" caused by the spreading of oceanic plates along a ridge in the Atlantic Ocean. An elastic rebound effect also comes into play, whereby the crust slowly adjusts to a new equilibrium following the removal of the substantial ice sheets that depressed the crust until approximately 10,000 years ago. Imagine pushing your hand into a sofa and then taking it away; your hand is like the ice sheet and the springy sofa is like the crust, except that the processes play out over very different time and space scales. The combination of these forces and the failed rift structure has conspired to produce higher levels of earthquake activity at New Madrid than elsewhere in the midcontinent—including temblors of prodigious size and impact.

The Earthquakes

Heralded by neither prophecy nor obvious physical signs, the New Madrid seismic zone sprang to life in the span of a heartbeat at 2:15 in the morning

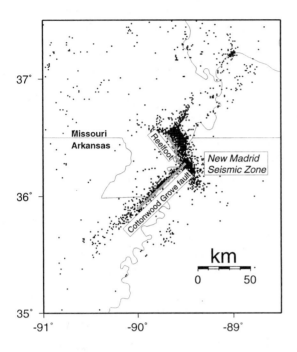

FIGURE 4.4. New Madrid seismic zone: circles indicate small earthquakes between 1974 and 1994; principal faults are also shown.

on December 16, 1811. Earthquakes were scarcely unknown in the region. As recounted by contemporary observer Daniel Drake, who left one of the more thorough accounts of the effects of the 1811–1812 events, significant temblors had rocked parts of the region in the decades prior to 1811: in 1776, 1791 or 1792, 1795, and 1804. These shocks were felt most strongly in locations from southern Illinois to near Lake Michigan, with a fifth significant event occurring near Niagara Falls, New York, in 1796.

Yet nothing could have prepared early settlers for what was to follow. From the depths of their slumber on a cold December night, residents of the region were abruptly awakened to a nightmarish scene. Furniture rattled and fell; chimneys crashed to the ground; in some cases walls followed suit. Loose soils in the Mississippi River Valley were shaken until they lost their internal cohesion and behaved like liquids rather than solids. Sand erupted from the ground via the phenomenon known as liquefaction, not just over the immediate epicenter of the earthquake but over a swath many tens of kilometers long. Disturbances within the Mississippi River caused banks to slump and fail for some distance along the river, in some cases carrying with them tangled mats of trees uprooted wholesale.

FIGURE 4.5. Early woodcut depicting scene on the Mississippi River during the New Madrid earthquake sequence. (State Historical Society of Missouri, Columbia, Mo.)

Boatmen on the Mississippi River that night described a scene of unimaginable terror and chaos (figure 4.5). Some boats—and, presumably boatmen—were lost; others felt blessed to have escaped with their lives.

And it was only the beginning. Local residents reported substantial aftershock activity throughout the remainder of the night; at dawn, a large aftershock occurred, nearly rivaling the main shock in severity. Yet even this second powerful event was but a taste of things to come. The aftershock sequence continued for many weeks, not starting to subside until after the middle of January 1812. Then, near 9 o'clock on the morning of January 23, another powerful earthquake struck, widely described as rivaling the first event in its strength. While the dawn event on December 16 was considered an aftershock, the event of January 23 was large enough to be considered a distinct new main shock in its own right (see sidebar 4.1). After another two weeks of energetic aftershock activity, the great sequence was punctuated by what local residents termed the "hard shock"[8] at 3:45 on the morning of February 7, 1812. Aftershocks would continue to rock the area for many months, with felt events continuing for several years, but the February event was the third

SIDEBAR 4.1

The conventional interpretation of the January 23, 1812, New Madrid main shock places the event at the northern end of the New Madrid seismic zone. One eyewitness, Eliza Bryan, described the event as being severe in the town of New Madrid. However, a recent study by the authors and colleague Karl Mueller reexamined this temblor, which has always been the most enigmatic of the principal 1811–1812 shocks. This study concluded that the location of the January shock cannot be determined with any precision from the available accounts. Notably, very few accounts describe actual damage during this shock; witnesses well to the north and northeast of New Madrid describe it as particularly strong. The new study proposed a possible new source zone for the earthquake: in southern Illinois, fully 120 kilometers north of the New Madrid region. A detailed account of the sequence describes prodigious effects in this location: massive sand blows and a two-mile-long "crack" across which two feet of movement occurred. Although this new study does not prove that the January 23, 1812, temblor struck in southern Illinois, it does show that the location of this event is highly uncertain. It might not have been a "New Madrid main shock" at all!

and last of the New Madrid main shocks. (Indeed, scientists generally consider current earthquake activity at New Madrid to be a continuing aftershock sequence—nearly two centuries after the main shocks!)

Even to earthquake-savvy present-day residents of California, the New Madrid sequence defies imagination: a devastating earthquake followed by a relentless aftershock sequence. Then another main shock. And then another.

In the early 1800s, the midcontinent was a sparsely populated place. But tens of thousands of hardy souls had ventured to the edge of the western frontier, and far more—hundreds of thousands—had settled throughout Tennessee, Kentucky, and Ohio. Many of these early settlers were witness to earth science history, but only a small handful recorded their observations in a manner amenable to preservation over the ages. Those who did so, earned a measure of immortality for their efforts. They are, and always will be, the small chorus of voices from the past; the ones whose words have become the only seismographs we will ever have, because the New Madrid sequence predates the development of modern mechanical seismometers by nearly a century. For such "preinstrumental" events, earthquake scientists rely on what are known as "felt reports"—the anecdotal accounts of shaking severity and its effects on people and structures—to determine magnitude and, in some cases, fault geometry. As discussed in chapter 1, such accounts are used to determine so-called intensity data for historic earthquakes.

For the New Madrid sequence, intensity data are vexingly sparse (figure 4.6), but they are by far the most direct information available from which the magnitude of the earthquakes can be evaluated. One should pause to listen carefully to those voices from the past. Most of what we can learn about the 1811–1812 earthquakes, we will learn from them.

Sources of earthquake information from the early 1800s generally fall into one of two categories: newspaper articles or letters. The former bear little resemblance to newspaper articles of today. With, typically, only four pages to cover national, state, and local news (and advertisements), most accounts of earthquakes span scarcely a single brief paragraph, such as the following example from Frankfort, Kentucky:

> On Monday morning last, at about half past two o'clock, the shock of an earthquake was very sensibly felt in this place, which lasted about two minutes. A little after three o'clock there were two more shocks, but neither of them as severe as the first. It shook houses so as to cause a very considerable alarm lest they should fall, and some bricks were thrown off the tops of chimneys [*sic*] by it.[9]

Letters, on the other hand—some of which were published in their entirety in newspapers—often are rich in detail and eloquent in their narrative. Accounts from boatmen on the river portray a tumultuous scene. One singularly sober account by John Bradbury begins:

> I was awakened by a most tremendous noise, accompanied by so violent agitation of the boat that it appeared in danger of unsettling. Before I could quit the bed, or rather the skin, upon which I lay, the four men who slept in the other cabin rushed in, and cried out in the greatest terror, "O mon Dieu! Monsieur Bradbury, qu'est ce qu'il y a?" I passed them with some difficulty, and ran to the door of the cabin, where I could distinctly see the river agitated as if by a storm; I could distinctly hear the crash of falling trees, and the screaming of the wild fowl on the river.[10]

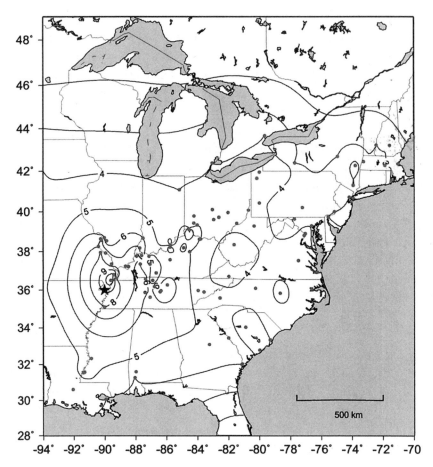

FIGURE 4.6. Map showing the distribution of shaking severity during the first New Madrid earthquake. Contour lines enclose regions of intensities as indicated on each line. Dots indicate locations for which eyewitness accounts exist.

Elsewhere along the river, Firmin La Roche described similar phenomena: "When I could see, the trees on the shore were falling down and great masses of earth tumbled into the river."[11] Even closer to the epicenter of the December event, William Leigh Pierce wrote of more spectacular sights:

> Here the earth, river, &c. torn with furious convulsions, opened in huge trenches, whose deep jaws were instantaneously closed; there through a thousand vents sulphurous streams gushed from its very bowels, leaving vast and almost unfathomable caverns. Every where nature itself seemed tottering on the verge of dissolution. . . . During

the day there was, with little intermission, a continued series of shocks, attended with innumerable explosions like rolling of thunder; the bed of the river was incessantly disturbed, and the water boiled severly [*sic*] in every part; I consider ourselves as having been in the greatest danger from the numerous instances of boiling directly under our boat; fortunately for us, however, they were not attended with eruptions. One of the spouts which we had seen rising under the boat would inevitably have sunk it, and probably have blown it into a thousand fragments.[12]

Although some of the phenomena described by Pierce—notably the "vast caverns"—are dubious, his account generally rings true. The waters of the Mississippi would not have boiled, of course, but the combination of river and riverbed disturbances could easily have generated a similarly tumultuous effect.

All told, some half-dozen boatmen recorded their experiences for posterity. Within the river towns, a comparable number of individuals left detailed written accounts. One important, and oft-quoted, letter was written by New Madrid resident Eliza Bryan to evangelist Lorenzo Dow, who published it in his journal. Bryan's letter begins:

On the 16th of December, 1811, about 2 o'clock a.m., a violent shock of an earthquake, accompanied by a very awful noise, resembling loud, distant thunder, but hoarse and vibrating, followed by complete saturation of the atmosphere with sulphurous vapor occurring [in] total darkness.[13]

A second contemporary account of the earthquakes found its way into print in the *Lexington Reporter*. Published under the headline "Extracts from a Letter to a Gentleman in Lexington, from His Friend at N. Madrid (U.L.) Date 16th Dec. 1811," the account was not ascribed to its author and has been referred to as the "New Madrid Extract." In their detail, their careful chronology, and their immediacy, these words carry us back to that fateful day with a singular clarity:

About two o'clock this morning we were awakened by a most tremulous noise, while the house danced about, and seemed as if it would fall on our heads. I soon conjectured the cause of our trouble, and cried out it was an Earthquake, and for the family to leave the house, which we found very difficult to do, owing to its rolling and jostling about. The shock was soon over, and no injury was sustained, except the loss of the chimney, and the exposure of my family to the cold of the night. At the time of this shock the heavens were very clear and serene, not a breath of air flirting: but in five minutes it became very dark, and a vapour which seemed to impregnate the atmosphere, had a disagreeable smell and produced a difficulty of respiration. I knew not how to account for this at the time, but when I saw in the morning the situation of my neighbors' houses, all of them more or less injured, I attributed it to dust and sut [sic] &c which arose from their fall. The darkness continued until daybreak; during this time we had eight more shocks, none of them as violent as the first.

At half past six o'clock in the morning it cleared up, and believing the danger to be over I left home, to see what injury had been experienced by my neighbors. A few minutes after my departure there was another shock, extremely violent—I hurried home as fast as I could, but the agitation of the earth was so great that it was with much difficulty I kept my balance—The motion of the earth was about twelve inches to and fro. I cannot give an accurate description of this moment; the earth seemed convulsed—the houses shook very much—chimnies [sic] falling in every direction. The loud hoarse roaring which attended the Earthquake, together with the cries, screams, and yells of the people, seem still ringing in my ears.

Fifteen minutes after 7 o'clock we had another shock. This was the most severe one we have yet had—the darkness returned and the noise was remarkably loud. The first motions of the earth were similar to the preceding shocks, but before it receded we rebounded up and down, and it was with difficulty we kept our seats. At this instant I expected a dreadful catastrophe—the uproar among the people heightened the coloring of the picture—the screams and yells were heard at great distance.[14]

Although the quakes are known collectively as the New Madrid sequence, modern estimates of the epicenter of the first event place it some 60 miles southwest of the town. Had the modern convention of naming earthquakes for the town closest to the epicenter been followed, they would have likely become known as the Little Prairie earthquakes, for the small village of Little Prairie was some 48 kilometers downriver of New Madrid, significantly closer to the locations of the December events. Its greater proximity to the earthquakes was reflected clearly in the level of damage it sustained. Whereas the "New Madrid Extract" spoke of toppling chimneys, the account of Mississippi River boatman James McBride, published in the *Quarterly Publication of the Historical and Philosophical Society of New York*, told of far greater devastation to the modest collection of houses in Little Prairie: "Of about a dozen houses and cabins [*sic*] which I saw, not one was standing, all was [*sic*] either entirely prostrated or nearly overturned and wrecked in a miserable manner; the surface of the ground cracked and fractured in every direction."[15]

Damage in the town of New Madrid itself was very much less severe, the dramatic nature of the shaking notwithstanding. The *Farmer's Repository* of January 31, 1812, included as part of a different account, "The only brick chimney in that place was entirely damaged by the shocks." It seems unlikely that an observer would have remarked on the demise of a chimney had the entire house been in ruins. However, damage at Little Prairie was clearly substantial, and McBride went on to describe dramatic disturbances to the earth itself:

> The surface of the ground was cracked in almost every direction and stood like yawning gulphs, so wide that I could scarcely leap over them, at other places I came to spaces of ground several poles in width, sunk down two or three feet below the common level of the ground. But what particularly attracted my attention were circular holes in the earth from five or six to thirty feet in diameter, the depth corresponding with the diameter so as to be about half as deep as wide, and surrounded by a circle of sand two or three feet deep, and a black substance like stone coal but lighter, probably carbonized wood. I took some pieces of this to the boat, and putting them on the fire I found they would burn, at the same time producing a strong and disagreeable sulphurous smell. These holes I presume must have been produced by a strong current of air issuing from the bowels of the earth, throwing up sand and water and this black substance which was perhaps wood, long imbedded in the earth, prostrating the trees and everything else where they happened and producing the most horrible disorder.[16]

McBride was right about *what*, but not about *why*. Today, earth scientists recognize the phenomenon of sand and water eruption as liquefaction, and the circular features as *sand blows*. Sand blows are created not by currents of air, but by prolonged and severe shaking that increases the pressure within buried layers of sand until they have nowhere to go but up. And mentions of a "sulphurous smell" are remarkably pervasive in accounts written both near to and far from New Madrid. Many witnesses also commented on the appearance of a "vapor" in the air. Although these accounts in some ways remain perplexing, they can generally be understood as a consequence of disruption to shallow sediment layers that are rich in organic matter in various stages of decomposition.

The damage that McBride witnessed represented the effect of not one earthquake but of multiple events, including the three severe quakes described in the "New Madrid Extract." Documentation from residents of Little Prairie reveals that the damage was caused not by the first event, but by one of the large shocks later in the morning.

While the "New Madrid Extract" is clear in its description of three large events, with the latter two roughly 45 minutes apart, accounts from more distant locales consistently report one especially severe aftershock at approximately 7:15 in the morning. Because severity of earthquake shaking depends on proximity to the event as well as magnitude, one can conjecture that the earlier of the two large aftershocks was felt especially strongly near New Madrid because its epicenter was very close to the town.

In any case, the 7:15 A.M. aftershock clearly seemed to produce greater shaking at both New Madrid and Little Prairie than did the 2:15 A.M. main shock, even though the latter is inferred to have been larger. Modern scientists have used these observations, along with others, to investigate the earthquakes in some detail.

Another detailed account, this one from approximately 16 kilometers east of Little Prairie, in Tennessee, describes the portentous effects of the 7:15 A.M. aftershock in particular. The author of this account, which was published in Samuel Cummings's *The Western Pilot* in 1847, did not identify himself by name, although he does give his age as being "scarce fifteen years old."[17] But, from information contained in the account, he was almost certainly John Hardeman Walker, who eventually returned to Little Prairie and went on to (literally) shape the future of the state of Missouri (sidebar 4.2).

In mid-December 1811, Walker was hunting with a companion, Jean Baptiste Zebon. They had traveled ten miles east from Little Prairie to find a lake that an Indian had described to Zebon, one that Walker found to be "as full of beaver and otter, as any lake I had ever seen."[18] He also described it as crescent-shaped and about three miles long.

Unlike some settlers, who left the New Madrid region never to return, John Hardeman Walker did return to the Little Prairie region after the earthquakes. By 1818, Walker was 25 years old and a successful cattleman. When the territory of Missouri applied to be admitted into the Union, the initial southern boundary was a simple parallel at 36.5 degrees north. This boundary would have put Walker's property in the territory of Arkansas, which at the time was only poorly organized. Walker and his neighbors lobbied and, as a result, Missouri got its boot heel. The modern joke in Arkansas is that by altering the boundary, the United States simultaneously raised the average IQ in both Arkansas and Missouri. The modern joke in Missouri is, of course, similar but reversed.

On the second night of the hunting expedition, Walker tells of being awakened by "a noise like distant thunder, and a trembling of the earth."[19] The shaking was hard enough to generate waves in the lake that Walker heard hitting the bank, and to shake the trees strongly enough to break off limbs. Frightening as this experience was, it paled in comparison with what was to follow at dawn:

> It was awful! Like the other — first, a noise in the west, like heavy thunder, then the earth came rolling towards us, like a wave on the ocean, in long seas, not less than fifteen feet high. The tops of the largest sycamores bending as if they were coming to the ground — again, one rises as it were to re-instate, and bending the other way, it breaks in twain, and comes to the ground with a tremendous crash.[20]

Even allowing for a measure of hyperbole in the account, the description of trees being broken implies ground motions at the upper range of shaking generated by earthquakes. But even that was just the beginning:

> Now the scene became awful in the extreme. Trees were falling in every direction — some torn up by their roots, others breaking off above ground, and limbs and branches of all sizes flying about us, and the earth opening, as it were, to receive us, in gaps sometimes fifteen feet

wide, then it would close with the wave. The water of our little lake was fast emptying itself in these openings, and as they would close, it would spout high in the air.[21]

The temblor did indeed spell the end of the lake, for a note at the end of Walker's account remarks that the former lake later become a beautiful prairie, "as high and dry a piece of ground as there is anywhere in the vicinity."[22]

Walker describes an arduous journey home, one that required careful navigation around fallen trees and cracks in the earth. Upon their arriving at the Mississippi River at sundown, Walker describes his friend's face turning, "pale as death" at the sight of their hometown deserted save for a solitary cow, lowing pitiably. With no way to cross, Walker and Zebon spent a sleepless night on the east side of the river, staying in constant motion to keep from freezing in the cold winter air. In the morning a man appeared in the village. Zebon hailed him, but Walker wrote that the man "ran from us as if the wolves were after him." As Zebon and Walker contemplated their fate, another man appeared, and launched a canoe into the river toward the beleaguered pair. This individual proved to be none other than Walker's father. He told them that the inhabitants of Little Prairie had left the village, a "mass of ruins," for an encampment two miles into the prairie. They had selected one of their group to return to report on the condition of the village, and this individual had returned breathless, telling of seeing a "ghost or the devil" on the opposite side of the river. Realizing that his son and his son's companion were likely the ghost(s), Walker's father had then returned to the village.

Within a few days of December 16, the good citizens of Little Prairie had clearly had quite enough. According to George Roddell, a respected senior resident of the village, the entire population of the town—almost 100 souls— stumbled into New Madrid on Christmas Eve, having heard that "the upper country was not much damaged."[23]

Those who fled from Little Prairie made the proverbial journey from frying pan to fire, for the earthquakes migrated north as well. Descriptions of a substantial new event on the morning of January 23, 1812, are sparse and scarce compared with those for the December main shock, but it was felt as least as widely as the latter had been.

While the January 23 event was clearly a substantial temblor, contemporary accounts from the New Madrid region are mostly notable for their absence of detail or elaboration. Perhaps the quake-weary residents felt disinclined to comment at length on events that were simply more of the "same old thing" by then.

As if the temblors weren't enough to contend with, a cold spell settled into the Mississippi River Valley in January, freezing the river. The ice began to break up only on January 22, opening the river to boats that had been waiting in and around Louisville, Kentucky. Thus, while few boats were on the Mississippi for the January 23 event, the river was well populated with boats and boatmen in the wee hours of the morning on February 7, 1812. The earth had set the stage in other ways as well. Eliza Bryan wrote of the ground being in a state of "continual agitation" from January 23 until February 4, when the area was shaken by another event that nearly rivaled the earlier ones. Four similarly strong shocks struck on February 5 and, finally, at approximately 3:15 on the morning of February 7, the final climax. In the words of Eliza Bryan, "a concussion took place so much more violent than those preceding it that it is denominated the 'hard shock.'"[24] Another observer, Robert McCoy, reported, "The [shock] that did material injury to the village of New Madrid was not until the 7th of February."[25] The definition of "material injury" apparently was not entirely clear, since other accounts of the earlier shocks certainly describe damage to the town. But while some residents had left and others had moved into light wood encampments, some semblance of a town had always remained. Those who arrived in New Madrid later on February 7, however, found the town all but deserted. The hard shock clearly led not just to evacuation, but to a flight of sheer terror. Among the "possessions" left behind by one family was a teenage daughter, Betsy Masters, injured in the earthquake and unable to join the exodus. An unrelated individual who was part of the exodus, Col. John Shaw, returned to Miss Masters's home shortly afterward and "cooked up some food for her, and made her condition as comfortable as circumstances would allow."[26] Beyond that, the fate of the unfortunate young woman remains unknown.

One also cannot but wonder after the fate of the 100 beleaguered souls who had made their way from Little Prairie to New Madrid in late December. Did they remain in New Madrid and experience the January 23 shock? Were they around to witness yet another earthquake large enough and close enough to leave catastrophic devastation in its wake? According to Walker's account, the mostly French citizens of Little Prairie continued northward from New Madrid to St. Louis, with the greater part of them continuing on to Canada, whence they had originally come. But apparently at least some of them—Walker included—stayed in the area, eventually returning to their homesteads in Little Prairie.

In general, many of the most dramatic accounts of the hard shock came not from the land, but from boatmen along the river, from 19–32 kilometers north of New Madrid to some 96 kilometers south. The experiences of a Kentucky boatman 19 kilometers north of New Madrid were relayed by William Shaler:

> He was awakened by a tremendous roaring noise, felt his vessel violently shaken, and observed the trees over the bank falling in every direction, and agitated like reeds on a windy day, and many sparks of fire emitted from the earth. He immediately cut his cable and put off into the middle of the river, where he soon found the current changed, and the boat hurried up, for the space of a minute, with the velocity of the swiftest horse; he was obliged to hold his hand to his head to keep his hat on. On the current's running its natural course, which it did gradually, he continued to proceed down the river, and at about daylight he came to a most terrible fall, which, he thinks, was at least six feet perpendicular, extending across the river, and about a half a mile wide.[27]

Mathias Speed, who had not been far away, chronicled similar experiences in a letter written to the *Pennsylvania Gazette*:

> We took the right hand channel of the river of this island, and having reached within about half a mile of the lower end of the town, we were affrightened with the appearance of a dreadful rapid of falls just below us. . . . As we passed the point on the left hand below the island, the banks and trees were rapidly falling in. From the state of alarm I was in at this time, I cannot pretend to be correct as to the length or height of the falls; but my impression is, that they were about equal to the rapids of the Ohio.[28]

Venturing into town the next day, Speed encountered a bleak scene: "There was scarcely a house left entire—some wholly prostrated, others unroofed and not a chimney standing—the people all having deserted their habitations, were in camps and tents back of the town."[29]

In many respects, rivermen's accounts of the hard shock appear to describe effects different in severity but not in quality from those of the earlier shocks. Yet the eye of the trained earth science observer immediately lights upon one particular detail: the appearance of "falls" across the Mississippi. During the earlier events, almost all of the disruption, or deformation, of the ground described is consistent with liquefaction effects and massive slumping of riverbanks and sandbars. When riverbanks slump, obviously they slump *toward* the river. But the creation of a waterfall across a river suggests something else—vertical movement on a fault. Walker's account from Tennessee also suggests vertical fault motion, for the 7:15 aftershock apparently created a dry and elevated prairie where a substantial lake had once been.

To contemporary observers, these phenomena were likely less striking; they would have been just two of many substantial disturbances along and adjacent to the river. The falls, moreover, proved to be short-lived. Under the continued onslaught of normal river currents, disruptions in the underlying riverbed were soon worn smooth. When the 1999 ChiChi, Taiwan, earthquake created a dramatic waterfall across a river, the ground displaced by fault movement was much more solid, and so the fall withstood the force of the river for some time (figure 4.7). The Mississippi River Valley, however, is not a setting conducive to preservation of geologic features.

The hard shock altered more than the ground beneath the Mississippi River; for a time, it altered the very flow of the river. New Orleans merchant Vincent Nolte wrote, "The current of the Mississippi . . . was driven back upon its source with the greatest velocity for several hours, in consequence of an elevation of its bed."[30] Today the Mississippi is often preceded by the word "mighty"; it is sobering to realize that the hard shock of 1812 was powerful enough to reverse its course. Some changes are beyond the power of even great earthquakes to affect permanently, however. Nolte went on to say:

> But this noble river was not thus to be stayed in its course. Its accumulated waters came booming on, and o'ertopping the barrier thus suddenly raised, carried before them with resistless power. Boats, then floating on its surface, shot down the declivity like an arrow from a bow, amid roaring billows and the wildest commotion.[31]

Mathias Speed observed his downriver progress to have been only four miles in the several hours following the shock. Between this realization and his observation of water rushing into the river from the adjacent woods, he concluded, astutely, "It is evident that the earth at this place, or below, had been raised so high as to stop the progress of the river."[32]

FIGURE 4.7. Waterfall created when the 1999 ChiChi, Taiwan, earthquake ruptured across a river. A similar situation is thought to have created waterfalls on the Mississippi River during the last large New Madrid earthquake. (Courtesy of National Oceanic and Atmospheric Administration/National Geophysical Data Center)

The violent convulsions of the river and riverbed led, not surprisingly, to dramatic subsequent effects. For some distance, the combination of shaking and river/riverbed disturbances caused banks to give way, in many cases wreaking havoc with the trees that had formerly flourished at water's edge. The Kentucky boatman 20 kilometers above New Madrid "observed whole forests on each bank fall prostrate, like soldiers grounding their arms at the word of a command."[33]

A century after 1812, geologist Myron Fuller compiled a map showing tracts of land that had been sunk, elevated, and otherwise disrupted by the earthquakes. Fuller documented one particularly dramatic effect of the earthquakes: the creation of a new lake—Reelfoot Lake—where none had existed before (figure 4.8). The land along Reelfoot Creek had been low and swampy prior to the February main shock, but afterward became a bona fide lake up to 6 or 7 meters deep. Entire groves of trees were submerged, eventually to wither to stumps that stood for decades as silent, ghostly sentinels. As documented by Fuller in 1912, "the line between the uplifted and submerged lands

FIGURE 4.8. Reelfoot Lake, created during the last large New Madrid earthquake. (Photograph by Eugene Schweig, used with permission)

at the south end of the lake is so sharp that it suggests faulting or at least a very sharp flexure."[34] It would be many more decades before scientists further developed this prescient observation.

The New Madrid earthquakes rearranged the landscape in significant and complex ways: lands were raised and sunk, trees toppled, new lakes were created and others were destroyed. The geologic handiwork of the New Madrid sequence provides a second body of evidence with which scientists can attempt to unravel the events. The February shock in particular left behind geologic testimony substantial enough to allow interpretation with at least some measure of confidence, even almost two centuries later.

Although many important questions remain unanswered, the closing years of the 20th century witnessed tremendous strides in our understanding of the complex geology of the New Madrid seismic zone and the enigmatic earthquake sequence of 1811–1812. Working on the assumption that current earthquakes are long-lived aftershocks, scientists have used sophisticated techniques to determine earthquake locations precisely; these then illuminate the faults that ruptured almost 200 years ago. The geometry of faults has been investigated using other methods as well, including detailed studies of topography and the underwater shape (bathymetry) of Reelfoot Lake that have shed further light not only on the geometry of the faults, but also on their movement in recent earthquakes. Together with careful analysis of historic accounts, this work has led to an identification—literally, a subsurface mapping—of the dipping Reelfoot fault that ruptured in the February 7, 1812, hard shock.

Scientists have teased out more results as well, including identification of the faults that ruptured in the December 16, 1811, main shock and dawn aftershock. And using modern earthquakes as calibration events, scientists have been able to estimate the magnitudes of the main shocks and their largest aftershocks (see sidebar 4.3).

The scientific story of the New Madrid earthquakes is remarkable not only for its ending but also for the epic journey that brought us here. When the New Madrid earthquakes struck, the United States was barely more than a collection of hardscrabble colonies and the field of modern instrumental seismology was a century away from existence. Moreover, these earthquakes occurred within the muck of the Mississippi embayment, with predictably disorderly and complex results. The journey—the remarkable series of scientific investigations to find faults in the midcontinent—has been the adventure of several lifetimes. This one earthquake sequence has taught scientists much of what we know about large earthquakes in unlikely places. Indeed, the 1811–1812 temblors still rank among the largest and best-studied earthquakes that have occurred away from active plate boundaries.

Modern scientists owe a tremendous debt to the individuals who witnessed the New Madrid earthquakes and endeavored to document their effects. These include people who wrote letters and a few remarkable individuals who sought to compile information and make sense of it. Chief among the latter group was Congressman Samuel Mitchell, who set out to collect and synthesize accounts of the temblors throughout the country and to understand what they implied for various theories of how and why earthquakes occur. Mitchell certainly brought considerable acumen to the task. Inevitably, however, he failed in his quest to understand earthquakes, because he lacked understanding of the broader scientific framework within which answers could be found. In 1815, only a sorcerer could have divined the complex nature of faults in the New Madrid region and understood the na-

SIDEBAR 4.3

The magnitudes of the principal New Madrid earthquakes have been the subject of considerable debate in the earth science community, with credible investigations estimating values as high as 8.0+ and as low as 7.0. For earthquakes of this vintage, eyewitness accounts provide the most direct information with which scientists can estimate magnitude. To make these estimates, recent earthquakes, for which seismometer data are available, are used as calibration events. This process is straightforward in theory but complicated in practice if there is no calibration event as large as the historic earthquake one is studying. The preferred estimates of one of the authors peg the largest New Madrid earthquake, on February 7, 1812, at magnitude 7.4. This value is lower than earlier estimates because the study took into account the detailed settlement patterns in early 19th-century America—in particular, the fact that most people lived immediately adjacent to waterways, where earthquake shaking is known to be amplified. Although the magnitudes will likely continue to be debated by the scientific community, scientists agree that earlier estimates of 8.75 are now outdated and erroneously high. Unfortunately, these earlier high estimates were responsible for the conclusion that the New Madrid earthquakes were the largest historic temblors in the contiguous United States—a conclusion that has evolved into folklore that, seemingly, refuses to die. And as long as we're on the subject of mythology, the New Madrid earthquakes did *not* ring church bells in Boston. A systematic archival search of Boston-area papers by seismologist Ronald Street led to the conclusion that the shocks were not felt in Boston at all.

ture of the forces that conspire to produce earthquakes there. But Mitchell was one in a long line of naturalists and scholars who endeavored to record basic data that they could not hope to understand, but nonetheless recognized to be of critical importance. Only nearly two centuries after Mitchell's publication was an entire generation of scientists able to exploit the data to their fullest potential. For a scientist there is no greater professional gratification than that derived from starting with a seemingly inscrutable collection of observations and, finally, understanding how all of the puzzle pieces fit together: finding faults in the New Madrid seismic zone and, moreover, understanding the intricacies of their dance over time.

Modern earth scientists are fortunate to be investigating the New Madrid earthquakes and region at a time when the answers are, at long last, within our grasp. We are not only happy but also forever grateful to the generations of astute observers and scientifically inclined souls who came before us, and whose legacy we are proud to honor.

The Historical Journey

If 18th-century Europeans increasingly perceived their world in rational, scientific terms, they were also increasingly aware of a larger world than the one their forebears had known. By the dawn of the 18th century, Europeans found themselves with a hitherto unimagined option to trade the Old World for a brand new one, far away and across a very big sea. The trade cannot have been made lightly, for the journey was an arduous one and a return journey not to be counted on.

Late 18th- and early 19th-century immigrants to the New World would have thus brought not only their cultural awareness and experiences, but also, clearly, substantial intestinal fortitude (literal and otherwise). Once in the nascent United States, those settlers who ventured farther westward still were likely cut from an especially sturdy cloth. The frontier beckoned not only with opportunities but also with heavy-duty challenges. Native Americans, disease, hunger, isolation—no shortage of perils to life and limb in the wilderness. Early 19th-century frontier settlers cannot have expected life to be easy, and it wasn't.

Yet people pressed westward. By 1812, the city of St. Louis had a population of about 1,400. Brackenridge tallied 12 mercantile establishments in the town, as well as a printing office. According to historian William Foley, the private libraries of St. Louis boasted some 2,000–3,000 different titles—sizable holdings for the day—at the time of the Louisiana Purchase. Considering St. Louis's sophistication and fortunate location, Brackenridge concluded that the city was destined to become "the Memphis of the American Nile." (In 1812 no city—only the Army's Fort Pickering—existed at the location of present-day Memphis, Tennessee.)

The New Madrid earthquakes rocked St. Louis hard enough to cause alarm and light damage to brick buildings, but the effects of the temblors were less severe here than in smaller towns closer to New Madrid. Hardest hit, of course, were the settlements closest to the earthquakes, including New Madrid and Little Prairie.

As discussed earlier, the 1811–1812 sequence was so relentless that even hardy frontier settlers staged a strategic retreat—in some cases all the way back to Canada. The towns of New Madrid and Little Prairie must have resembled ghost towns in the immediate aftermath of the sequence. Who in their right mind would choose to stay there? The frontier had proven itself hostile even beyond settlers' imaginations; the earth itself had been unwelcoming with a vengeance.

One resident of Kentucky, Joseph Finklin, expressed his sentiments about westward expansion in a letter to Samuel Mitchell:

The Indians cannot have suffered much in their tents and bark houses. But the United States will suffer in the sales of their public lands west of the Mississippi for an age. At least the present generation must be buried before the spirit of wandering, in that direction, revives; and may it not be an advantage that some power exists to fix a boundary for our fellow citizens; for my own part, I am pleased in viewing the benefits which my country will derive from this great shock.[35]

In the aftermath of the New Madrid sequence, the U.S. Congress passed the New Madrid Claims Act—the first disaster relief bill in the United States. The intent of this act was to allow settlers with damaged land to trade their New Madrid claim for an equivalent amount of land elsewhere. This act, which might have cemented New Madrid's demise, was, by most accounts, an unmitigated failure, spawning fraudulent claims and land speculation by the unscrupulous. At least some settlers, however, were able to relocate under the terms of the act.

Even so, Finklin's forecast was not to be borne out. Yes, New Madrid itself struggled; but according to Brackenridge and others, the town was scarcely thriving before the earthquakes. And according to Amos Parker's 1835 *Trip to the West and Texas*, New Madrid was "slowly regaining its former condition" (such as it was!) by 1834. Parker noted that the ponds created by the earthquake still rendered the area unhealthy, yet went on to observe that "New Madrid is, however, quite a village, transacts much business and is the most noted landing place for steamboats on the west side of the river below St. Louis."[36]

In general, an examination of early settlement patterns reveals that while the very earliest towns clung tenaciously to the riverbanks that were both their lifeblood and their curse, settlements tended to migrate landward, by a few to a few tens of kilometers, as the 19th century wore on. According to historian William Foley, as of 1804 few people except Indian traders had ventured more than a few miles into the Louisiana Territory. However, according to historian James Shortridge, by 1823 many settlers had migrated inland from rivers by a few to as much as 32 kilometers.

And westward expansion continued apace. The population of St. Louis more than tripled between 1811 and 1820. By 1850 it was pushing six figures. In the same time period the population of Missouri grew from just over 20,000 to approximately 66,000.

The New Madrid earthquakes must have contributed to early settlers' perceptions of the frontier as a hostile place, and of riverbanks as especially inhospitable. The latter recognition can only have grown over time, as more people experienced more winters in the midcontinent—and the capacity of the large rivers to cause large floods. The sinuous Mississippi River would have inevitably revealed its fickle and hazardous nature. The powerful New Madrid earthquakes of 1811–1812 were only one in a collection of hazards, none of which proved menacing enough to stem the growing tide of westward expansion.

If the earthquakes did not scare people away, one might imagine that the temblors at least frightened them into revising any previously sinful ways. With a paucity of scientific answers, earthquakes were still frequently viewed as a reflection of the hand of God (God or the Great Spirit, depending on one's cultural tradition). A prolonged sequence of powerful temblors cannot but have seemed like a dramatic and personal message for settlers whose church attendance had been less than exemplary. As summarized by James Finley in his autobiography, "It was a time of great terror to sinners."[37]

A careful examination of church membership statistics shows that the tumult of 1811–1812 did have an impact on settlers' churchgoing habits. Historian Walter Posey focused on the membership of the Western Conference of the Methodist Church, which in 1811 included Tennessee, Kentucky, and parts of Mississippi, Arkansas, Illinois, Indiana, Ohio, and West Virginia. Although the total population of this region grew by only a few percentage points between 1811 and 1812, the membership of the Western Conference ballooned by over 50 percent, from just over 30,000 to almost 46,000 devout souls.[38]

Clearly the New Madrid earthquakes put the fear of God into the hearts of many settlers. Whether or not the sequence had a lasting impact on religious practices, however, is far less clear. Over the five years following 1812, the Western Conference grew by fewer than 5,000 persons, a far slower rate than during the years prior to 1811 or subsequent to 1818. Perhaps, then, the earthquakes did not really save souls that would have otherwise been lost; perhaps they only succeeded in bringing them into the flock a little sooner.

In the annals of the New Madrid sequence one finds another interesting footnote concerning religion. According to an account that first appeared in the Bedford, Pennsylvania, *Gazette* in 1814, the December main shock caused the residents of Louisville, Kentucky, to "[grow] very devout in one night."[39] On the next day, and with long faces, "they subscribed a thousand dollars to build a house of public worship." No further action was taken until the January main shock, at which point another thousand dollars was raised. And again the matter rested (the account continues) until the February main shock, at which point a third sum of a thousand dollars was collected. When the earthquakes did not return, the people of Louisville "concluded the devil would not send for them for a few years more,"[40] and decided to use the $3,000 for the construction of a fine theater.

The account makes for a fun story, but, according to a later comment in the Louisville *Western Courier*, the theater in question had in fact been built prior to the start of the New Madrid sequence. One learns to take New Madrid anecdotes with a grain of salt.

The Louisville story resonates, however, because, at first blush, it fits all of our preconceptions. No atheists in foxholes—not many in recent earthquake zones, either. But the punch line to the story resonates as well; the joke works because of the basic truth on which it is built. Earthquakes rock us in a spiritual as well as a literal sense. A prolonged sequence like New Madrid will, moreover, take an especially hard toll on the psyche. Yet even this powerful sequence did not have a significant impact on larger societal and cultural trends. The earthquakes not only failed to stop westward expansion; they scarcely amounted to as much as a speed bump. Early American settlers were not so easily scared off. A few of them abandoned land that probably would have been abandoned soon anyway; of those who moved, most did not go far. And within a small handful of years, the frontier nation was bustling and booming—but only in an economic sense—once again.

So complete was the rebound that by the mid-20th century, the New Madrid sequence was scarcely a footnote in earth science circles—and not even a footnote in the public awareness of earthquake hazard. Only in the closing decades of the 20th century did scientists' attention return to the sequence, and eventually drag public attention back as well. In retrospect one can scarcely imagine a scenario that would have had a more lasting societal impact than a prolonged sequence of large earthquakes striking the very heart of a newly established frontier. And yet the impact was minimal at best. Frontiersmen (and -women) are, it seems, made of sterner stuff.

5

19th-Century Temblors:
A Science Is Born

When the observer first enters upon one of those earthquake-shaken towns, he finds himself in the midst of utter confusion. The eye is bewildered by "a city become an heap." He wanders over masses of dislocated stone and mortar, with timbers half buried, prostrate, or standing stark up against the light, and is appalled by spectacles of desolation. . . . Houses seem to have been precipitated to the ground in every direction of azimuth.
> —Robert Mallet, in Charles Davison, *The Founders of Seismology* (Cambridge: Cambridge University Press, 1927), 76

When the 1811–1812 New Madrid earthquakes rocked the North American midcontinent, the state of seismology as a modern field of inquiry matched that of the United States as a country: a few steps beyond a collection of struggling colonies but still very much a work in progress. Inevitably the New Madrid earthquakes caught the attention of some of the best and most scientifically trained individuals who experienced the remarkable sequence. Jared Brooks, Daniel Drake, Samuel Mitchell—all endeavored to document their observations, and other data, in a thorough and scientific manner, and all had some awareness of theories of the day. Yet in retrospect, contemporary investigation of the sequence remained limited and fragmented. The response would be very different when the Charleston, South Carolina, earthquake rocked the central United States toward the end of the 19th century. Although still too early to be recorded on seismometers, the Charleston earthquake was investigated in a manner that the present-day earth scientist would consider modern.

Within the United States, the New Madrid sequence and the Charleston earthquake form a pair of 19th-century bookends. On the shelf in between these milestone events was the emergence of seismology as a modern science—developments that, for the most part, took place outside of the United States. The field would not mature fully during the span of this single century, of course. Elastic rebound theory, the fundamental tenet that has lent its name to this book, would not appear as a fully realized theory until shortly after the 1906 San Francisco earthquake. Plate tectonics, which at long last explained the engine behind the earthquake machine, would arrive on the scene later still. But it was during the 19th century that scientists first began to build the very foundations of the field. As earlier chapters have discussed, intelligent speculations about earthquakes date back at least as far as Aristotle's time; intelligent, thorough observations of earthquakes date back to the mid-18th century, in particular the efforts sparked by the Year of the Earthquakes in England (1750) and the Lisbon earthquake five years later.

What transpired during the 19th century is this: for the first time, scientists began to establish some of what we know today about earthquakes and the waves they generate. For a nascent field of scientific inquiry to take shape, initial developments must be made by scientists who do not consider themselves experts in the particular field—almost by definition, since the field does not exist. Thus early developments in seismology fell to individuals with training in other disciplines, frequently engineering. Seismology is, after all, largely a science of mechanics: Who better to begin to understand its rules than individuals with an aptitude for and training in mechanical systems?

Initial 18th-century strides in seismology to a large extent involved the simplest observational characterization of earthquakes and earthquake sequences: where and when they had happened in the past. In the 19th century, scientists (and engineers) began to probe deeper into the science of earthquakes. In so doing, pioneering scientists set into place the first solid stones upon which a fully mature science could later be built. This chapter is about a century of scientific craftsmanship: the earthquakes that inspired it, the craftsmen who carved sound and stalwart foundation blocks from hitherto unfashioned stone, and, finally, the foundation blocks themselves.

The Great Neapolitan Earthquake

Among the attractions that beckon visitors to Italy today are the country's surviving medieval villages, small hamlets where time seems to stand still (figure 5.1). Perched on mountaintops, which offered both defensible sites and an escape from sometimes unhealthy low-lying valleys, these villages today resemble nothing so much as a fairy tale: a snapshot of a distant, simpler, and (at least viewed through the rose-colored glasses of retrospection) more romantic time.

From a seismological point of view, however, these villages resemble nothing so much as disasters in the making. It does not require a recent degree in either seismology or engineering to arrive at this conclusion. In his 1905 book *A Study of Recent Earthquakes*, Charles Davison wrote, "Less favourable conditions for withstanding a great shock are seldom, indeed, to be found than those possessed by the mediaeval towns and villages of the meizoseismal area."[1] By "meizoseismal area," he meant the area that had been impacted by the 1857 earthquake. His observation continued:

> In buildings of every class, the walls are very thick and consist as a rule of a coarse, short-bedded, ill-laid rubble masonry, without thorough bonding and connected by mortar of slender cohesion. The floors are made of planks coated with a layer of concrete from six to eight inches thick, the whole weighing from sixty to a hundred pounds per square foot. Only a little less heavy are the roofs, which are covered with thick tiles secured, except at the ridges, by their own weight alone. Thus, for the most part, the walls, floors, and roofs are extremely massive, while the connections of all to themselves and to each other are loose and imperfect.[2]

Davison also commented on the location of early towns and villages that they are "generally perched upon the summits and steep flanks of hills, especially of the lower spurs that skirt the great mountain ranges."[3] Robert Mallet—one of seismology's founding fathers, about whom the reader will

FIGURE 5.1. Picturesque Erice, Sicily, typifies the medieval villages still found throughout Italy. (Susan E. Hough)

hear (much) more shortly—observed that this rugged topography amplified earthquake shaking, a phenomenon confirmed by modern seismologists. Adding insult to injury—or, more specifically, injury to injury—the streets in such villages were traditionally very narrow, sometimes a scant five feet wide. When houses started to crumble and give way, they fell against and on top of each other. As pioneering geologist Deodat Dolomieu observed following the 1783 Calabria earthquake, "the ground was shaken down like ashes or sand laid upon a table."[4] The 1857 great Neapolitan earthquake, as it came to be known, took a heavy toll. Of a regional population of just over 200,000, nearly 10,000 lives were lost. At the village of Montemurro, the death toll reached a staggering 5,000 out of a population of only 7,002, with an additional 500 injured. At Saponara, almost half of the town's 4,010 residents were killed.

The tolls of the 1693 Sicilian and the 1783 Calabria earthquakes were greater still. As is typically the case, fatality estimates for these two temblors vary considerably between different sources. Some 60,000 lives are generally believed to have been lost during the 1693 shock. An earlier temblor in Calabria in 1638 was witnessed and documented by noted Jesuit priest and scholar Athanasius Kircher. (Among his myriad accomplishments, Father Kircher is credited with first introducing the term "electromagnetism" in a book published in 1641.) Approaching the gulf of Charybdis, Kircher observed that the waters "seemed whirled round in such a manner, as to form a vast hollow, merging to a point in the centre. Proceeding onward and turned my eyes to Aetna, I saw it cast forth large volumes of smoke, of mountainous sizes, which entirely covered the island."[5] Noting the unusual stillness of the water, Kircher warned his companions that an earthquake was on the way. (This according to his own account; however, one is inclined to take a Jesuit at his word.) His prediction was borne out shortly after the party arrived at the Jesuit college in Tropea. "The whole tract upon which we stood seemed to vibrate, as if we were in the scale of a balance that continued wavering. This motion, however, soon grew more violent; and being no longer able to keep my legs, I was thrown prostrate upon the ground. In the mean time, the universal ruin round me doubled my amazement."[6]

Leaving what he called the "seat of desolation," Kircher and his companions set sail to Rochetta, where they "saw the greater part of the town, and the inn which we had put up, dashed to the ground, and burying the inhabitants beneath the ruins." At a distance of some 60 miles, Stromboli continued "belching forth flames in an unusual manner, and with a noise which I could distinctly hear."[7] Another violent temblor struck while Kircher's party was in Rochetta.

Kircher also tells of finding a young boy along the shore whom they asked for information. The young man, however, appeared "stupefied with terror," unable to find a voice, let alone answers to the questions asked him. When

Kircher's group tried to offer him food and comfort, "he only pointed to the place of the city, like one out of his senses; and then running up into the woods, was never heard of after. Such was the fate of the city of Euphaemia."[8] Journeying some 200 miles [320 kilometers] along the coast, Kircher found "nothing but the remains of cities; and men scattered without a habitation, over the fields."[9]

The 1638 temblor appeared to be an isolated event, but both the 1693 and the 1783 earthquakes were in fact a series of strong shocks—as is all too typical for earthquake activity in Italy. The latter sequence included six strong shocks and many more smaller ones over a two-month period, causing enormous destruction and claiming some 50,000 lives. This series of temblors was investigated by the first approved "earthquake commission," and kindled some of the earliest efforts to build modern earthquake recording devices in Italy. Today, Italy remains at the forefront of modern earthquake science, a situation borne out of sad exigency: no country in Europe has lost as much in earthquakes.

Medieval villages are still sprinkled throughout the Italian countryside; many more modern-looking towns have grown up with medieval villages at their core. Houses in such villages remain as vulnerable as they were in Mallet's day. When the 1968 M5.4 earthquake and subsequent large aftershocks struck near Gibellina, Sicily, they reduced the town to heaps of rubble. In spite of the temblor's relatively modest magnitude, some 400 lives were lost in Gibellina and surrounding villages. Following this earthquake, authorities took the unusual and controversial step of entombing the rubble in concrete, preserving the former layout of blocks and streets to create a giant and ghostly jigsaw puzzle memorial (figure 5.2). Artist Alberto Burri "plastered over" the rubble to create a village-size sculpture called *Cretto*, which preserves not only the layout of the former village but also the memory of the tragic event that ended its existence. Wandering through the narrow "streets" of the memorialized Gibellina, one grasps a sense of scale that words and photographs strain to capture. Formerly a thriving village, now it is gone.

Italy is characterized by a markedly diffuse earthquake hazard: moderate-to-large earthquakes strike throughout the country, but any one area will see a damaging temblor only infrequently. Thus some medieval villages have survived to this day. But time is not on their side. All of these villages represent targets on a macabre dartboard, targets on which lethal darts will almost

FIGURE 5.2.   The formerly picturesque village of Gibellina, Sicily, was transformed into rubble by a earthquake in 1968, and later transformed into a village-sized memorial to the disaster. (Susan E. Hough)

certainly land someday. A magnitude 7.2 dart landed on Messina, Sicily, in 1908, killing 70,000–100,000 in the quake itself and subsequent tsunami (figure 5.3). The Messina quake struck early on the morning of December 28, the Monday after a Christmas weekend, and left both Messina and the main-land town of Reggio Calabria—combined populations of 210,000—in nearly total ruin. This temblor remains the deadliest in known European his-tory. The M7.5 Avezzana earthquake of 1915 claimed nearly 30,000 lives. The most severe shock in recent memory, an M7.2 shock, struck southern Italy in November 1980, killing 3,000 and leaving a staggering quarter-million people homeless in and around the regions of Campania and Basilicata.

Robert Mallet

The great Neapolitan earthquake of 1857 was neither the biggest nor the most destructive historic earthquake in Italy. According to recent investigations, the magnitude was approximately 7.0. But it stands apart in two respects: the extent to which it was investigated and the extent to which it sparked nascent theories in earthquake science. These distinctions can be credited almost entirely to the efforts of one man: Robert Mallet. Born in Dublin in 1810, Mallet entered Trinity College to study engineering at the age of 16. At 21 he joined his father's business, further developing the engineering capabilities of a factory that had previously made products called sanitary fittings and also small fire engines. According to Charles Davison, before long, "all engineering work of any consequence in Ireland was carried out by the firm."[10] Davison describes Mallet's "first great feat": raising the massive (133-ton) roof of St. George's Church in Dublin.

FIGURE 5.3. A postcard depicts damage and rescue operations following the devastating Messina, Italy, earthquake of 1908.

Mallet's interest in earthquakes had already been piqued by the time the Neapolitan temblor struck. Mallet secured a grant from the Council of the Royal Society and departed the following February for the kingdom of Naples. Armed with official letters of authority (the likes of which remain helpful in post-earthquake investigations), Mallet completed his survey of the region impacted by the earthquake—nearly all of southern Italy—in a scant two months. He returned home to complete what would prove to be a landmark report. Mallet viewed this report as not only a documentation of the earthquake but also, according to Davison, as something of a textbook of observational seismology. Many of the ideas that Mallet developed or tested, including methods to determine the origin point and depth of an earthquake, would not stand the test of time. Although mechanically reasonable, such methods were fatally flawed by virtue of an incomplete understanding of the nature of earthquake waves and how they originate.

Mallet also never stumbled on the fault that slipped in 1857. In fact it was not discovered until recently, when geologist Lucilla Benedetti and her colleagues undertook extensive field investigations and found a hillside scarred by earthquake ruptures. It is possible that Mallet failed to find the surface rupture because it was hard to find: he believed the disturbance was confined many miles underground. It is also possible the rupture was buried under snow. Had he noted the 16-kilometer crack through the countryside, it seems very likely that this mechanically inclined individual would have made something of the link between fault rupture and quakes.

Mallet's exemplary and thorough observations of the temblor's effects, however, do represent an enduring contribution—as does the precedent he established when he undertook the investigation and report. His map of the earthquake's effects (shown in figure 1.3) was not the first such map, but it was among the earliest, and arguably one of the most important. Mallet identified four levels of damage ranging from zone 1, some 65 by 37 kilometers, in which destruction was nearly total. Within zone 1 he identified the focus of the disturbance as a "fissure" 16 kilometers long, the production of which he surmised was accompanied by the injection of steam at high pressure. Here again, Mallet's mechanically sensible interpretation missed the mark, but one cannot denigrate the magnitude of the accomplishment.

At the low end of Mallet's scale he mapped out zone 4, which marked the extent of what he called the "disturbed area."[11] This zone included nearly the full southern half of Italy's boot—all except the very tips of the toe and heel.

Mallet's engineering sensibilities shine through his exposition on the earthquake and his interpretations of the damage. He can scarcely be faulted (so to speak) for not being correct in every inference, since his work preceded all but the most fundamental theories of earthquake waves. Still, Mallet's enduring contributions to the development of modern seismology were legion, including pioneering efforts to documents the speed and amplitude of waves. He also introduced several terms that remain familiar to the modern seismologist: isoseismal, seismic focus, and meizoseismal area, among others. He and his son put together the first global catalog of all earthquakes, an enormous endeavor spanning more than 3,000 years and nearly 7,000 earthquakes. When Robert Mallet died in 1881, he surely left the field of seismology on surer footing than when he found it.

Richard Oldham

As the 19th century drew to a close, northeastern India was rocked by a disastrous earthquake that was investigated by another Dublin-born seismologist: Richard Oldham. His father, Thomas Oldham, a professor of geology at Dublin University and director of the Geological Survey of Ireland, had been appointed the first director of the Geological Survey of India. The interests of the son followed those of the father, propelled by the need to collaborate on two important articles on Indian earthquakes started by the elder Oldham. The first was a historical account of all known earthquakes in India, which, like the earlier work of Mallet and his son, included verbatim extracts of all earthquake reports from the earliest times, starting in classical languages. The second article, completed after the death of his father, described in detail the effects the damaging 1869 earthquake in Cachar in northeastern India. It was this second article that prepared Richard Oldham for the 1897 M8 great Assam earthquake in the neighboring Shillong Plateau.

For both Oldhams, earthquakes in India were a very far cry from the shocks that had rocked their native Great Britain. Although a number of historic earthquakes in and around England were instrumental in inspiring investigations of earthquakes as a natural phenomenon, rarely are England and its staid environs visited by shaking of disastrous proportions. As a colonial power, however, England found herself faced with earthquakes on an entirely different scale—in particular in India, the northern fringes of which are now known to be an active zone of collision where the formerly separate

Indian plate continues to career into Asia. This collision creates monster mountains—the tallest peaks on the planet—and, to go along with them, monster earthquakes.

Earthquakes have exacted a heavy toll on India over the years, and continue to represent a hazard of potentially catastrophic proportions. When one speaks of rebound, nowhere as much as in India is one reminded of what is at stake. Earthquakes may not wipe out civilizations, but they certainly can wipe out villages, towns, cities—taking a great many lives as well as inflicting huge financial losses.

India's earthquake history has been documented, with varying degrees of completeness, as far back as the early 16th century. A great earthquake struck the Himalayas in 1505, collapsing the city of Agra, 96 kilometers to the south. In the early 19th century, the Allah Bund earthquake struck the western state of Gujarat, not far from the location of the devastating Bhuj earthquake of 2001.

One of the most remarkable earthquakes of the 1800s struck Assam, in northeastern India, as the 19th century was drawing to a close, on June 12,1897. It generated perceptible shaking throughout most of the Indian subcontinent. The zone of highest damage, in which the destruction of masonry buildings was almost total, extended over a radius of nearly 320 kilometers, covering a region roughly the size of England that was centered on a region known as the Shillong Plateau. Charles Davison quotes an eyewitness: "The ground began to rock violently, and in a few seconds it was impossible to stand upright, and I had to sit down suddenly on the road. The shock was of considerable duration, and maintained roughly the same amount of violence from the beginning to the end." This account continues, "The school building . . . began to shake at the first shock, and large slabs of plaster fell from the walls at once. A few moments afterwards the whole building was lying flat, the walls collapsed, and the corrugated iron roof lying bent and broken on the ground."[12]

Modern analyses yield a magnitude value of 8.0–8.1 for the 1897 temblor: a truly enormous earthquake, the likes of which are seldom seen outside of the planet's subduction zones. One might imagine this estimate to be imprecise by virtue of its antiquity, but in fact it is a notably precise estimate for an earthquake of its vintage. One might well wonder how scientists can know much about an earthquake that happened so long ago. The answer, in this case, is several answers. Perhaps most important, the great trigonometric survey of India had covered the Shillong Plateau in the 1860s, providing a reference baseline against which the changes caused by the earthquake could be measured, using a method known as triangulation. Following the earthquake, Captain John Bond was dispatched to locate and remeasure the original survey points, a formidable challenge in disease-ridden forests drenched by an annual rainfall of 3.3 meters. The data collected by Bond and his team revealed an astonishing uplift in some areas of over 7 meters, a vertical shift that is difficult to measure accurately with triangulation, and which was hence dismissed as too outlandish to be correct.

The more accurate horizontal triangulation revealed about 3.5 meters of relative motion in places, but Oldham recognized a fundamental ambiguity in interpreting these measurements. The results could be explained by two different scenarios: either the plateau had expanded as it lurched upward and northward, or it had lurched southward as a block. The interpretation of leveling data requires calculation of position shifts relative to one line that is held fixed in the calculation, but in this case the answer changed depending on which of the several dozen lines one held fixed. Oldham realized that a second survey along the northern edge of the plateau would resolve the ambiguity, but it was not conducted until 1936. When analyzed with modern methods and sensibilities, the new data confirmed that the plateau had shifted up and northward, sliding on a gigantic fracture dipping southward beneath the plateau (figure 5.4).

Further information about the earthquake is provided by a small number of very early seismogram recordings as well as the detailed distribution of damage and other effects. Again following in Robert Mallet's footsteps, Oldham set out to survey the effects of the temblor. His observations were summarized in one of seismology's more famous early isoseismal maps (figure 5.5). A century or so later, seismologist Nicholas Ambraseys conducted an exhaustive search for and analysis of original accounts of the event. This massive undertaking led to the creation of a somewhat different—and far more complex—view of the earthquake's effects. Using modern earthquakes as calibration events, analysis of these detailed data confirms the magnitude estimate determined from triangulation data.

Oldham's report—sometimes referred to as his great memoir—on the 1897 temblor was published by the Geological Survey of India in 1899. Like Mallet's report on the Neapolitan earthquake four decades earlier, Oldham's report provided two enduring contributions: the data he collected and his pioneering scientific interpretations. In addition to his isoseismal map, Oldham mapped surface rupture along the Chedrang fault, to the tune of al-

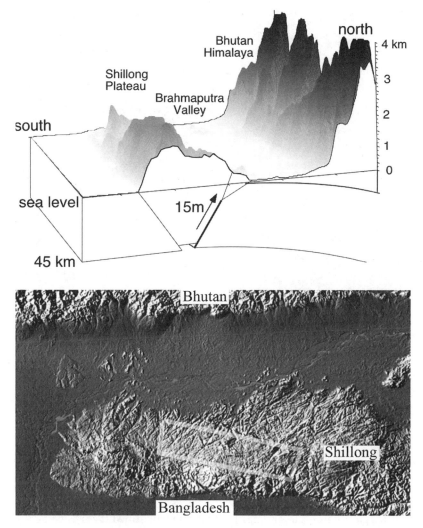

FIGURE 5.4. The powerful Assam earthquake caused further uplift of a region known as the Shillong Plateau. The earthquake ruptured along an angled fault, as shown in the top panel, and terminated far below the surface, leaving no primary surface break for geologists to find.

most 11 meters of movement at one location. He documented evidence that stones had been thrown into the air during the shaking, implying that vertical accelerations had been at least as strong as gravity. He further examined seismograms recorded at 11 stations between 6,880–7,840 kilometers from Shillong.[13] In these records Oldham found three distinct groups of promi-

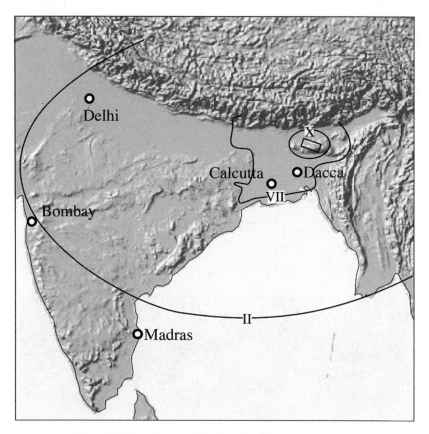

FIGURE 5.5. Richard Oldham's map of shaking distribution during the 1897 Assam earthquake. Later work by Nick Ambraseys and one of the authors revised this simple view considerably. Here II indicated extent of felt area; X indicates total collapse.

nent wiggles (figure 5.6) and, for the first time, identified them as being the compressional (P) and shear (S) waves generated by the earthquake, followed by the slower surface waves.

Oldham's extraordinary scientific acumen extended still further. In spite of having mapped the Chedrang fault, which had created waterfalls and lakes, and dammed rivers, Oldham speculated that the true seat of the disturbance had been more than 16 kilometers deep, along the type of thrust plane that had been mapped in the Scottish Highlands. As summarized by Davison, "A great movement along one of the main thrust-planes would carry with it dependent slips along many of the secondary planes. Direct effects of the former might be invisible at the surface, except in the horizontal dis-

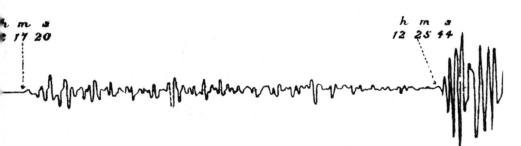

FIGURE 5.6. Early seismogram showing initial P wave (left) and later S wave (right). The seismogram was recorded by an early seismometer in Catania, Italy, following the 1897 Assam earthquake. (Plate XLI from Richard D. Oldham, "Report on the Great Earthquake of 12 June 1897." *Memoirs of the Geological Society of India.* Calcutta: Geological Survey of India, 1899)

placements that would be rendered manifest by a renewed trigonometrical survey; whereas the latter might or might not reach the surface, giving rise in the one case to fissures and fault-scarps, in the other to local changes of level, and in both to regions of instability resulting in numerous after-shocks."[14] This might read like so much scientific mumbo jumbo, but to the ears of a modern-day seismologist it echoes resoundingly with prescience, anticipating much of what we now know to be true about thrust faults, both those that reach the surface and those that scientists now know as "blind."

Oldham concluded that the great Assam earthquake had been caused by a thrust fault about 320 kilometers long, at least 80 kilometers wide, and at least 16 kilometers deep. This inference was based largely on human seismometers—Indian telegraph operators who pressed their keys at the exact moment they felt one of the many thousands of aftershocks. A century later, one of the authors of this volume, along with colleague Philip England, revisited the old triangulation data to infer that the earthquake had ruptured about 112 kilometers of the buried Shillong thrust fault; they estimated the ruptured fault extended about 40 kilometers deep into the earth's crust, with the upper edge of the rupture 8–10 kilometers below the surface. Oldham was thus off by a factor of 2–3 in length and of 2 in width, and dead-on in his estimate of depth. For someone who was, so to speak, flying quite blind without seismometers or modern analysis methods, one can only be astounded that his estimates came so close to the mark. Oddly, although Richard Oldham went on to discover the core of the earth using seismic waves, in conversation with the eminent seismologist Sir Harold Jeffreys he asserted that he was mainly a geologist and not really interested in seismology. Sir Harold is on record as stating that "He [Oldham] was the only man I have known that did first-rate work in a subject that did not interest him."[15]

As Oldham documented and later researchers confirmed, the Assam earthquake was one of the most widely felt events in history. Although its magnitude did not rival those of the world's greatest subduction zone earthquakes, it was a massive event—also one whose waves traveled especially efficiently through the old and cold crust of the Indian subcontinent. Interestingly—and mercifully—the death toll did not reflect the nature of the earthquake. According to the best modern estimates, some 1,500 people were killed. This surprisingly low figure reflects several factors, including the sparse population density of the hardest-hit region as well as the relative safety of small, traditional Indian dwellings.

Ironically and potentially tragically, India's explosive population growth has not only placed more people in harm's way, but also has placed them in harm's way in considerably more dangerous structures. Small, traditional thatched-roof dwellings have given way to larger residential structures, built for the most part from vulnerable and often low-quality materials. Rebound from future great Himalayan earthquakes may thus be far less elastic than in the past.

But this is a subject for another chapter. Notwithstanding nontrivial societal impacts, the enduring legacy of the 1857 and 1897 temblors is largely one of scientific discovery. Mallet and Oldham were scarcely the only two 19th-century seismologists to make enduring contributions. But they surely stand tall among the field of pioneering scientists and engineers who crafted seismology into a field of modern inquiry—not quite starting from scratch, but certainly starting with precious few fundamental building blocks already in place. They may not be household names on a par with Pasteur, Curie, or Maxwell, but within earth science circles they are without question among the earliest founding fathers. Their legacies not only endure, but continue to grow every time scientists take a new look at the data they collected, and use these dusty but venerable observations to push the edge of the envelope a bit further still.

6

The 1886 Charleston, South Carolina, Earthquake

Four years later in 1890, the only visible evidence of this great
distruction [*sic*] was seen in the cracks which remained in
buildings that were not destroyed. A new and more beautiful,
more finished city had sprang [*sic*] up in the ruins of the old.
—Paul Pinckney, letter in *San Francisco Chronicle*,
May 6, 1906

Setting the Stage for Disaster

By 1886 the population of the United States had grown to over 50 million
people. Both the East Coast and the Midwest were by this time well populated with bustling towns and cities. Railroads had sprung up as well, greatly
facilitating land travel, which in turn helped spark further migration and
trade. The tide of westward expansion had long since steamrolled over whatever reservations the New Madrid earthquakes might have caused.

By 1886 the gold rush was already several decades old, and San Francisco
had grown into a lively urban center with a population of 35,000—about
5,000 more than the population of Chicago. A number of notable earthquakes had occurred in California by the end of the 19th century. While the

massive Fort Tejon earthquake of 1857 occurred too early in the state's history to leave a lasting impression on the collective psyche, large earthquakes along the eastern Sierras in 1872 and on the Hayward fault in 1868 had begun to suggest that California might be earthquake country.

Still, as of the late 1800s people had nothing approaching a modern understanding of earthquakes—neither their underlying physical processes nor their fundamental characteristics. As the 19th century drew to a close, scientists did not have any way to gauge the overall size of an earthquake, for scales had been developed only to rank the severity of shaking from a particular earthquake at a particular location. Whereas scientists today can easily rank temblors in terms of their overall size, or energy release, in earlier times people could only gauge an earthquake's overall effects, an assessment that can sometimes prove misleading. For example, the overall reach of earthquake shaking depends on the nature of the rocks through which the waves travel. As noted in chapter 5, waves travel especially efficiently in central and eastern North America, and especially inefficiently in California. Thus an earthquake of a given magnitude will pack a disproportionately heavy punch in the former region.

Moreover, the effect of earthquakes is not limited to the severity of shaking per se. The secondary effects, such as landsliding, slumping of ground along coastlines, and rock falls, can depend on the nature of the terrain as much as on the earthquakes themselves. Thus, by virtue of having created such enormous disruptions—in part for having occurred along the Mississippi River Valley—the New Madrid earthquakes appeared to trump anything that California had yet meted out. Damaging earthquakes had, moreover, occurred in New England in the 18th century. In their awareness that large earthquakes could happen almost anywhere, 19th-century Americans would have possibly been a step ahead of many Americans today, who are inclined to see earthquakes as a "California problem."

South Carolina was home to about a million people by the time the 1886 earthquake struck. It was originally explored by John Cabot in 1497, and the name Caroline—later Carolina—was bestowed by a group of 26 French Huguenots who established a settlement about 80 kilometers southwest of Charleston in 1562. These settlers later returned to France. The first permanent settlement of people of European descent began in Beaufort in 1670 under a grant from King Charles II of England. Within about a decade the original colony had moved to the present location at Charles Town, a settlement that thrived and developed into an important port.

By 1886 the streets of Charleston had grown into a web stretching over the full extent of the narrow peninsula formed by the convergence of the Cooper River to the south and the Ashley River to the north. Early maps reveal a network of prominent creeks fingering their way into the peninsula (figure 6.1); as the city grew, many of these creeks were filled so that they could be used as building sites. Well before 1886 builders recognized that deep foundation pilings were necessary to ensure sound construction in such areas, although it is extremely doubtful that anyone had earthquakes in mind.

Although there were relatively few government or public buildings in Charleston in 1886, the city was home to several substantial churches—notably St. Philip's and St. Michael's, as well as large, solidly built private res-

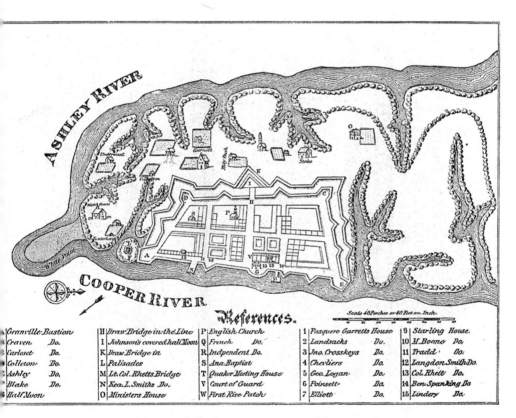

FIGURE 6.1. Map of the city of Charleston, 1704 . Many of the river inlets were later filled in to create more buildable space in the city. (Plate VIII from Clarence Edward Dutton, "The Charleston Earthquake of August 31, 1886." *Ninth Annual Report of the United States Geological Survey, 1887–88.* Washington, D.C.: Government Printing Office, 1889)

FIGURE 6.2. Solidly built Charleston house dating to the pre-Revolutionary period. (Figure 2 from Clarence Edward Dutton, "The Charleston Earthquake of August 31, 1886." *Ninth Annual Report of the United States Geological Survey, 1887–88*. Washington, D.C.: Government Printing Office, 1889)

idences dating back to Revolutionary times (figure 6.2). A fire in 1838 destroyed a large number of the wood houses in the city; the conflagration was so devastating to the city that a law was passed outlawing wood construction. Bricklayers, aware of the demand for substantial numbers of new homes in a hurry, arrived from the North, and brought with them new ideas about the best methods for bonding bricks. The strength of any brick structure depends critically on the quality of mortar and overall workmanship. The new

methods did produce strong walls; unfortunately, they also provided opportunities for shoddy work by lax or poorly supervised workers.

Charleston's history could scarcely have left the city less well equipped
to withstand the earthquake that was in its future. The most solidly built
masonry construction remains vulnerable to earthquake shaking; such
structures lack the overall flexibility and structural strength of wood-frame
buildings. As the citizens of Charleston were to find out, even the best-built
walls—"[rivaling] the masonry of old Rome in their solidity,"[1] according to
Dr. Gabriel E. Manigault, could not withstand the assault of a powerful
earthquake.

One can scarcely fault 19th-century residents for their lack of concern for
earthquake preparedness. While scientists know that earthquakes can strike
virtually anywhere, even today they grapple with the question of why coastal
South Carolina has been an unusually earthquake-prone region throughout
recent geologic times. In the 19th century people might have known that
earthquakes were possible anywhere, but previous large earthquakes in the
area had occurred well before historic times.

That small earthquakes were experienced infrequently in South Carolina
was also not particularly notable: by the closing years of the 19th century,
small, and even not-so-small, temblors had been documented in many parts
of the United States. Accounts of earthquakes were, however, published in
South Carolina newspapers virtually as soon as the first newspapers appeared. Temblors did not strike the region commonly, but often enough: 16
earthquakes were reported between 1698 and 1879. None was big enough to
cause damage beyond broken windows and crockery, but several, including
one on January 8, 1817, rattled hundreds of miles of the Atlantic seaboard.

A small burst of foreshocks that preceded the 1886 earthquake was therefore not entirely without precedent, but in retrospect they were quite noteworthy in one respect: their energetic rate of occurrence. Whereas prior to
this time earthquakes had occurred as singlets, with no more than three
shocks in any given year, 1886 was, in retrospect, a busy year. According to Dr.
Manigault, several light tremors were felt in the early summer of 1886, but
only as faint rumbles that did not receive much attention. Only after the fact
did people look back and wonder at the significance of these barely perceptible rumblings. The early summer temblors were felt, or at least reported,
only in Charleston, but at least three shocks between August 27 and August
29 were strong enough to be distinctly felt in the village of Summerville,
some 30 kilometers to the northwest of Charleston. At 8:00 o'clock on the

morning of August 27, residents of the village of Summerville heard a blast that many took to be a cannon or shot. The sound was described as "loud, sudden, and startling,"[2] accompanied by a single jolt of the ground. Residents considered the possibility that it might be an earthquake but, from what they had read of the phenomenon, were inclined to dismiss the explanation.

Indeed, one might well ask how it is that people *hear* earthquakes, when earthquake waves travel through the ground. The answer is that P waves are sound waves, and while their tones are generally too low to be heard by humans, earthquakes can sometimes generate sound waves in the audible frequency range. Like small musical instruments, small earthquakes produce relatively high tones—in physics parlance, this corresponds to high frequencies. Although this energy is damped out before it travels very far in the earth, the waves from small, nearby earthquakes can be heard as anything from a rumble to a boom. The energy travels through the ground, of course, but can be transmitted into the air. The low tones produced by large earthquakes are not generally within the audible range for humans, but a very large earthquake can produce a lot of energy over a wide range of tones, and thus can be heard as well. (When an earthquake is large enough, distinguishing any sounds from the earth itself can be difficult in the midst of the cacophony generated by manmade structures under extreme duress.)

The small but jarring temblors of August 27 were not followed by any further perceptible events that day, but the earth reasserted itself around 5:00 o'clock the following morning, in a stronger shock that abruptly wakened sleepers and alarmed those already awake. This temblor was followed by a succession of further light tremors throughout the day, some of which were accompanied by sounds. Residents of Summerville generally came to believe that they had in fact experienced earthquakes, but in general the reports generated little concern. Newspapers in Charleston, where nothing more than slight tremors had been felt, reported the accounts from Summerville, but conveyed the impression that perhaps the earthquakes were nothing more than some mundane event transformed by overactive imaginations.

In any case, the earth quieted down through August 30 and during the day on August 31. Whatever had come had apparently gone. By the night of August 31 any excitement or anxiety had likely subsided, and many residents of Charleston and its neighbors had, as was the fashion at the time, retired early for the night. Summerville resident Thomas Turner, president of the Charleston Gas Light Company, described the evening as "unusually sultry, but clear, and beautifully starlight."[3] Turner was in his garden enjoying the beauty of the late summer evening when, "without any rumble or warning, the floor seemed to sink under me."[4] From this instant Turner struggled, mostly unsuccessfully, to keep his footing as a series of jolts threw him backward and forward. Turner's sister-in-law, who had been entering her room as the tumult began, was also thrown onto the floor, and could only manage to crawl into the hallway. Amid the roaring and furious motion of the house an oil lamp tipped, starting a fire that Turner and his family were able to extinguish with carpet pieces as the violence of the shaking subsided. At least one large subsequent jolt then struck, tossing family members from side to side in hallways as they struggled to leave the house amid a sea of broken masonry and plaster.

When a fault begins to fail in a large earthquake, two things happen. First, the actual fault motion, or slip, moves along the fault at a high rate of speed, and earthquake waves are radiated into the surrounding rock from the initial point of failure, and then, in turn, from every point along the fault that moves. An observer at some distance from a large earthquake will thus experience shaking generated along the entire length of the fault as the rupture speeds along—just as a stationary observer will hear an extended train whistle from a train moving along a track. Moreover, a moving fault generates both P and S waves along its full extent, and while the two wave types quickly produce a jumble of energy at sites away from the fault, the initial motion felt by any observer is virtually always the P wave, which is almost always smaller than the S wave. There is a point to this seismological digression: for a number reasons, shaking from earthquakes—even large ones—commonly begins relatively inconspicuously before building to more severe levels. Turner's account, however, describes a marked absence of preliminary tremors, suggesting that he was standing immediately atop the part of the fault that began to fail. His description of a sensation of falling, rather than an abrupt jolt, further suggests that the beginning of the earthquake was relatively gradual. The onsets of large earthquakes are known to vary considerably, with some beginning relatively abruptly and others being relatively slow to pick up steam.

A sluggish onset of the earthquake is also consistent with accounts from Charleston, where a number of detailed accounts echo that of Carl McKinley, a member of the editorial staff of the *Charleston News and Courier* newspaper, who was in his second-floor office at the time of the main shock. McKinley's account begins as follows:

The writer's attention was vaguely attracted by a sound that seemed to come from the office below, and was supposed for a moment to be caused by the rapid rolling of a heavy body, as an iron safe or a heavily laden truck, over the floor. Accompanying the sound there as a perceptible tremor of the building, not more marked, however, than would be caused by the passage of a car or dray along the street. For perhaps two or three seconds the occurrence excited no surprise or comment. Then by swift degrees, or all at once—it is difficult to say which—the sound deepened in volume, the tremor became more decided, the ear caught the rattle of window-sashes, gas-fixtures, and other movable objects; the men in the office, with perhaps a simultaneous flash of recollection of the disturbance of the Friday before at Summerville, glanced hurriedly at each other and sprang to their feet with the started question and answer, "What was that?" "An earthquake!" And then all was bewilderment and confusion.[5]

McKinley's eloquent account continues:

The long roll deepened and spread into an awful roar, that seemed to pervade at once the troubled earth and the still air above and around. The tremor was now a rude, rapid quiver, that agitated the whole lofty, strong-walled building as though it were being shaken—shaken by the hand of an immeasurable power, with intent to tear its joints asunder and scatter its stones and bricks abroad, as a tree casts its over-ripened fruit before the breath of the gale.[6]

McKinley goes on to describe the crashing of stone and brick and mortar, accompanied by a "terrible roar [that] filled the ears and seemed to fill the mind and heart." Then, a fleeting second or two of respite before the violence increased once again to levels as severe as before. Fearing for their lives—and doubting their ability to make it from the building to safety outdoors, McKinley and his coworkers rode out the concussion until they felt the "blessed relief of . . . stillness,"[7] and then rushed down the stairway and onto the street. Through the feebly illuminated darkness of night a whitish cloud of dry, choking dust arose—the detritus of mortar and masonry reduced to rubble. Men and women, some barely dressed, rushed into streets littered with piles of bricks and telegraph wires dangling from broken poles. McKinley described crowds passing by—and not stopping to investigate—a woman lying motionless under a gas lamp. Nor did the crowds take heed of a man who walked along the sidewalk, his clothing soaked with blood from a head wound.

The scene took on the surreal aura of a nightmare: "The reality seems strangely unreal; and through it all is felt instinctively the presence of continuing, imminent danger, which will not allow you to collect your thoughts or do aught but turn from one new object to another."[8] The primal nature of terror and shock in the face of mortal danger—and the focus on immediate survival at the expense of more considered response—could scarcely be better summarized.

And on the heels of the first disaster, the second inevitably followed: bright lights appearing suddenly to further illuminate the ghostly scene, the shouts of "Fire!" quickly erupting from the crowd. As the fires began to ignite and spread, a strong aftershock struck, about eight minutes after the main shock— probably the same one that the Turner family experienced in Summerville: "the mysterious reverberations swell and roll along like some infernal drumbeat summoning them to die."[9] As the shaking and noises subsided, a new and more terrible din arose to take their place: shrieks of terror calling desperately for help amid the shattered buildings and growing fires.

Some 60 lives were lost that night, from a population of about 50,000 (figure 6.3). The injured were tended by doctors who worked heroically through the night. Those beyond help were laid on the ground in parks and public squares, covered only by shawls or sheets. Strong aftershocks punctuated the precarious calm, with four severe shocks occurring before midnight and three others between 2:00 and 8:30 in the morning.

McKinley's detailed, almost lyrical account allows the modern scientific reader to draw a few conclusions about the earthquake. As noted, the initial, modest rumblings were likely generated by the first few seconds of fault rupture near Summerville. As the rupture picked up speed—developing into what seismologists term unstable rupture—it would have immediately begun to pump far more energy into the surrounding rock. Energy from the beginnings of the rupture would have taken a few seconds to reach Charleston; from this time on, the city would be jolted by a seamless succession of waves generated along the moving rupture. Worse yet, as we will discuss shortly, it is possible that Charleston's location left the city in the crosshairs, so to speak, of the moving fault rupture.

The earthquake itself—the actual fault motion—almost certainly lasted for no more than a minute, and probably closer to 30 seconds. Barely a hic-

FIGURE 6.3. Damage from the earthquake. (Plate XVII from Clarence Edward Dutton, "The Charleston Earthquake of August 31, 1886." *Ninth Annual Report of the United States Geological Survey, 1887–88*. Washington, D.C.: Government Printing Office, 1889)

cup in time, the enormous energies released during such a short interval stand in testimony to the staggering scale and forces of earthquakes compared with our usual humble, human-scale endeavors.

And as is so often the case, the worst of times brought out the best in people in Charleston on and after that tumultuous first night. Trains leaving the city were crowded with refugees from the stricken city; they were carried free of charge if they could not afford the fare, and were greeted with hospitality at their destinations. Within the city, public officials remained at their posts, pastors ministered to their congregations, firemen and hospital workers toiled long hours, and private citizens helped tend to those in need. Usual social barriers fell away; accounts tell of the more fortunate caring for the less fortunate, with "fortunate" defined according to immediate need rather than color of skin or social status. McKinley observed that the shared travails of the city "showed, as could not have been shown under any other circumstances, how strong is the tie that yet binds the races together."[10]

The city demonstrated remarkable resilience over the long haul as well. Within a small handful of years a revitalized, by most accounts more beautiful, city had sprung up from the ashes. For one business, the earthquake and subsequent recovery became an integral part of its corporate identity (figure 6.4). Faults returned to their usual state of slumber, producing only a modest smattering of aftershocks in subsequent years, and gradually the extraordinary events of 1886 became only a footnote in the city's long and storied history.

Among those who chronicled the aftermath of the 1886 temblor was Paul Pinckney, descendant of prominent Charleston builder Charles Pinckney and a man of singular misfortune as far earthquakes were concerned. Having experienced the 1886 earthquake as a boy, he later moved to San Francisco, a move that would give him the opportunity to compare and contrast

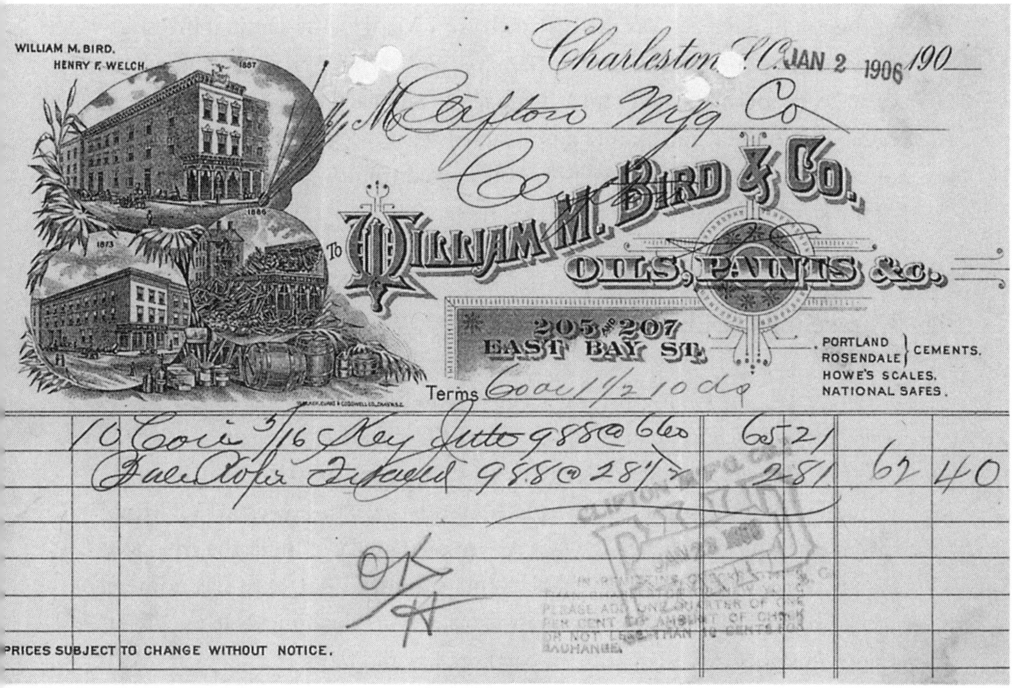

FIGURE 6.4. Sales invoice from William M. Bird & Company. For decades after the earthquake, the company's recovery from the temblor remained part of its corporate identity, with panels depicting the original company building, the wreck of the building following the earthquake, and the rebuilt structure.

two of the greatest earthquakes in American history. As some of his fellow San Franciscans despaired the fate of their ruined city, Pinckney wrote that they had clearly forgotten the lessons of Charleston: "Four years later in 1890, the only visible evidence of this great distruction [*sic*] was seen in the cracks which remained in buildings that were not destroyed. A new and more beautiful, more finished city had sprang [*sic*] up in the ruins of the old."[11] Cities do not, of course, spring up of their own volition, but the resources required to rebuild Charleston were modest in the scheme of things—not in the same league as the resources that would have been required to fully restore Lisbon in 1755. The earthquake clearly also galvanized a relatively prosperous citizenry to pick up and rebuild—if anything, better than before.

Thus was the broader societal impact of Charleston far more modest than that of more devastating temblors such as that of Lisbon and the great earthquake that would strike Japan in 1923. The social rebound of Charleston is fairly easily told. The rest of this chapter will focus on the scientific impact of the earthquake. As was the case with New Madrid, the seminal investigations of eyewitnesses and contemporary scientists have provided substantial grist for modern efforts to investigate faults and earthquakes in South Carolina. It is, once again, an intriguing tale of seismosleuthing—a marriage of old observations and new science. And so we turn now to the scientific part of the story.

Unraveling the Charleston Earthquake

Scientifically, the Charleston earthquake would have been a far bigger boon to earthquake science had it occurred just a few years later, when the world's first modern seismometers began to record the waves from large earthquakes, generally from the planet's active plate boundaries. The powerful 1891 Mino-Owari earthquake in Japan, the 1897 Assam earthquake in India, the great 1906 San Francisco earthquake—these events produced not only seismometer recordings but also long and dramatic surface scars that quickly led scientists to new insights into the nature of earthquakes and faults.

In the central and eastern United States, the 20th century would prove to be as quiet as the 19th century had been tumultuous. As the science of seismology came into its own, scientists increasingly recognized zones—typically narrow bands—of heightened earthquake activity in a number of places around the globe, even before the theory of plate tectonics provided an explanation for the observation. Like New Madrid, Charleston largely fell off the radar screens of scientists who investigated earthquakes, only to be redis-covered—and recognized as important—as the 20th century drew to a close.

Nevertheless, the reverberations—both literal and figurative—of the Charleston earthquake have played out quite differently over different time scales, from shaking of almost inconceivable violence in the short term, to relatively inconsequential aftereffects in the intermediate term, to the not insubstantial scientific reverberations in the long term. A scant 30 seconds of fault motion sparked decades of scientific inquiry aimed at investigating faults in South Carolina and understanding the hazard those faults pose.

As was the case at New Madrid, modern scientific techniques have been used to probe the crust in coastal South Carolina, with varying degrees of success. But, again echoing our experiences at New Madrid, much of what modern science knows about the Charleston earthquake has been gleaned from the written accounts of individuals who witnessed the events and/or documented their aftermath. Whereas the New Madrid accounts have been painstakingly assembled from disparate and widely scattered sources, ac-counts of the Charleston earthquake were assembled with painstaking care in the aftermath of the temblor, and published in 1889 in the Ninth Annual Report of the U.S. Geological Survey. Well known to earth scientists as the Dutton Report, this remarkable document was assembled by Capt. Clarence E. Dutton of the U.S. Ordnance Corps. Many of the accounts of the earthquake presented in this chapter were published as part of this report; one of Dut-ton's most judicious and important actions was to invite detailed accounts from a small number of witnesses who were known for their scientific acu-men and sensible temperament. However, the Dutton Report includes in-valuable information not just on the nature of the immediate effects in the epicentral region but a number of other types of information as well. These are highlighted in the following sections.

Clarence E. Dutton

Scientists pursue historic earthquake research because historic earthquakes are important. As noted previously, to understand earthquakes and earth-quake hazard, it is critical that we fully explore and exploit the data—albeit often imperfect—about earthquakes that occurred prior to 1900.

While earth scientists may dive into historic earthquake research intent on finding out more about earthquakes and the earth, they will almost inevitably learn other sorts of lessons along the way. Some educational tangents are obvious: clearly, one cannot hope to understand a historic earthquake unless one understands its historic context, including the language of the day. For instance, historic earthquakes in English-speaking countries are described in words appropriate for their day, not necessarily the same words that people would use in modern times.

More intriguingly, however, the historic earthquake scholar finds himself or herself learning something about historic figures. One doesn't have to read very many accounts of historic earthquakes before finding out that, more often than not, important historic events were chronicled in great detail by a small number of observers. Prior to 1900 the field of seismology was in its infancy, so such observers had little or no training, and usually limited prior awareness of earthquakes. Often, however, the individuals who step forward as impromptu seismologists have professions suggesting a certain scientific or technical bent. Three of the most valuable accounts of the New Madrid earthquakes were written by Samuel Mitchell, Jared Brooks, and Daniel Drake, respectively a congressman with training in geology, an engineer, and a physician. Typically, the observers are also highly accomplished; Drake's myriad contributions to medicine and Mitchill's government service testify to the extent of these men's industry and intellect.

Reading historic earthquake accounts, one is quickly reminded that human beings today may draw on a larger base of knowledge than our predecessors, but we are no smarter than the individuals who came before us. Another revelation follows: individuals such as Mitchill, Brooks, and Drake are impelled by intellect and curiosity to understand the spectacular natural events that unfurl before them, yet they are also doomed to failure. And nobody is as acutely aware of their inevitable failure as they themselves. In knowing more than most people, these sorts of individuals are also more keenly aware of what they do not know, of the limits of their intellect and understanding.

Thus historic earthquake research inevitably leads the researcher to reflections on, and appreciation for, the nature not only of the earth, but also of the human intellect. And perhaps nowhere can this aspect of historic earthquake research be better appreciated than within the pages of the Dutton Report. Although it represents a tangent to the central scientific themes of this book, the tangent is perhaps not inappropriate. Like any adventure worth its salt, scientists' investigations of historic earthquakes invariably involve surprises, discoveries as delightful as they are unexpected. Scientists understand this phenomenon and have a favorite word to describe it: serendipity. The fluke that nobody saw coming can be the one that leads a scientist in the direction that nobody imagined, and therein lies the very essence of discovery. In some cases, serendipity has a more human angle: the discovery that a particular individual is a fascinating story in his or her own right.

When Charleston was rocked by the great temblor on the night of August 31, 1886, Capt. Clarence E. Dutton of the U.S. Ordnance Corps was at the Warm Spring Indian Reservation in central Oregon. Because the earthquake predated modern wireless communications by over a century, Dutton did not learn of the earthquake until he reached Portland ten days later, and he did not return to Washington, D.C., for another three weeks.

Dutton may have been on the other side of the country when the earthquake occurred, but in every other respect he was the right man at the right time. Following service during the Civil War, he transferred to the Ordnance Corps in 1864. A graduate of Yale, Dutton pursued geology as an avocation after the war. Detailed to Washington, D.C., in 1871, he quickly cultivated relationships in scientific circles. In particular he sought out the acquaintance of scientists from the U.S. Geological Survey, including the institution's second director, Maj. John Wesley Powell. Dutton knew Powell, and Powell knew Ulysses S. Grant; as a result of these connections, Dutton was detailed—still as an Army officer—to the Geological Survey in 1875.

Over the following decade Dutton's efforts turned to fieldwork in the American West. He contributed to the development of the theory of isostasy, which describes how mountains remain in equilibrium by virtue of compensation in the underlying crust. A naturalist at heart, Dutton penned *The Tertiary History of the Grand Canyon District*, which is still admired for its scientific as well as its literary merits. He also carried out geologic investigations, with emphasis on volcanic processes, in California, Oregon, and Hawaii.

In letters that resonate in modern ears, in 1885 Dutton wrote that his home institution was slated to become part of an organized Department of Science, and voiced concerns that the enhanced bureaucracy would soon compromise the quality of science produced by the Geological Survey. Indeed, the dark forces of bureaucracy did succeed in prematurely ending Dutton's short but remarkable scientific career. In 1890 a new head of the Ordnance Corps looked askance at the 15-year-old arrangement that had allowed Dutton an annual furlough to pursue civilian work. No doubt some political and military leaders looked askance at Dutton in general. He was not only a man of boundless intellect, charm, and curiosity, but also a man who maintained an aura of irreverence—not, one imagines, the sort of personality type to endear him to those inclined toward military order and decorum. Dutton found himself detailed to the arsenal at San Antonio, Texas, where the isolation and intellectual vacuum were likely as stifling as the heat.

Fortunately for the world of seismology, the Charleston earthquake occurred in the 1880s and not the 1890s. By the time Dutton returned to the East Coast in the fall of 1886, other scientists had already been dispatched to the area: W. J. McGee from the Geological Survey, and Prof. Thomas C. Mendenhall of the U.S. Signal Service. In Charleston, McGee met a local resident, Earle Sloan, and, recognizing the man's keen intellect and predilection for scientific investigation, effectively deputized him to conduct a survey of the epicentral regions. Sloan mapped out the pattern of surface disruptions in impressive detail, identifying zones of cracked ground and bent railroad tracks. He also identified two primary "epicentrums"[12]: locations where, in his estimation, the effects were greatest (figure 6.5).

In the meantime, Ensign Everett Hayden from the Navy was charged with the task of assembling accounts of the earthquake throughout the country. This effort involved the mailing of circulars, the substance of which was also published in newspapers. This effort inspired a note of levity from one editor, who published his own humorous survey ("Did the photograph of your mother-in-law remain on the wall after everything else had fallen off?"[13]), but also garnered scores of serious, and valuable, responses.

Along with surveys both around Charleston and farther afield, the effort to document the earthquake included the solicitation of detailed accounts from two individuals who had experienced the temblor firsthand and who were judged to be particularly well suited to chronicle the event: Dr. Manigault, of the Charleston College, and Carl McKinley, the assistant editor of the *Charleston News and Courier*, both of whom we have heard from earlier in this chapter. Inevitably, compilation of the myriad surveys and accounts, along with analysis and discussion of the earthquake's effects within the context of basic physical principles, fell to the man of singular vision, intellect, and drive: Captain Clarence E. Dutton.

The Dutton Report

Mallet's pioneering treatise on the nature of earthquake waves had been published in 1846. In this and subsequent work by Mallet and others, some

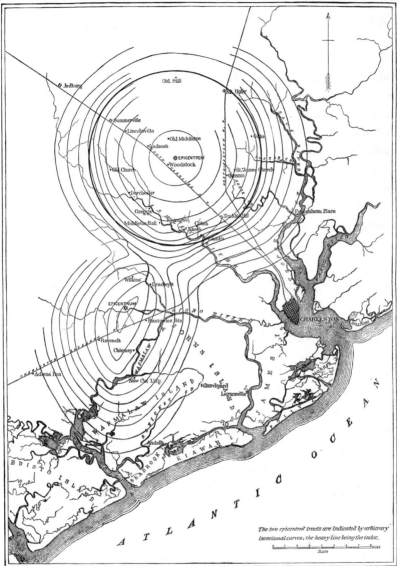

FIGURE 6.5. Charleston resident Earle Sloan surveyed the effects of the earthquake, shown in this remarkable map. Seismologists now believe that the earthquake began near Summerville, close to the "epicentrum" indicated near Woodstock. The initial rupture moved to the southeast along the Ashley River fault. A second rupture was then likely triggered on the Woodstock fault to the southwest. This fault is oriented roughly through the elongated contours around the second, southern "epicentrum" indicated on the map. (Plate XXVI from Clarence Edward Dutton, "The Charleston Earthquake of August 31, 1886." *Ninth Annual Report of the United States Geological Survey, 1887–88.* Washington, D.C.: Government Printing Office, 1889)

of the most basic tenets of earthquake wave theory were established. Scientists had not fully grasped the link between faults and earthquakes, but they had begun to make sense of the waves they generate. Although not right about everything, Mallet posited that earthquakes generate longitudinal waves that radiate in all directions away from an epicenter and can be analyzed by appropriately constructed mechanical devices. To measure the amplitude of vertical waves, he proposed a device involving a large mass (a cannonball) attached to the bottom of a spring, with a pipe attached to the bottom of the mass and dipping into a container of ink. Call it the dipstick seismometer: when the mass moved, the amplitude of the vertical motion would be revealed by the level of the ink on the pipe.

By the mid-1880s the science of seismography had progressed to the construction of the earliest instruments that scientists would now consider modern: devices capable of generating pen recordings of wiggles as the ground moved during earthquakes. Overwhelmingly, early theories in both seismology and seismography were articulated by scientists working in Europe and Japan. Geologists employed by the nascent U.S. Geological Survey focused their efforts on geologic mapping, as the "survey" part of the name implies. Although the New Madrid earthquakes presumably had not disappeared from the collective American consciousness by the late 1800s, earthquakes had not declared themselves to be an inordinate concern for the country in the decades following 1812. The large earthquakes occurred in the West during this time, including two noteworthy events in California: the M7.8 Fort Tejon and M7.6 Owens Valley quakes of 1857 and 1872 had not had much impact on the sparsely populated frontier state.

As much as any other single event, the Charleston earthquake introduced Americans to the science of seismology. And as much as any other single individual, Clarence Dutton was the man responsible for the introduction.

Following a detailed and exemplary compilation of the effects of the earthquake, in chapter 7 of the report Dutton proceeds to analysis of the observations. Chapter 7 is titled "The Speed of the Shocks" and is devoted to a remarkably careful analysis of the reported time of the initial shaking. Today seismologists can easily identify the onset of a P wave on a seismometer, usually to the nearest 1/100 of a second, and use such observations to learn how fast waves travel within the earth. Just a few seconds of reflection will reveal how much more difficult this exercise would have been in the absence of seismometer recordings—especially in the absence of the highly accurate and standardized clocks we have today.

Dutton amassed a list of over 200 accounts of the time of the shock at different locations. Some of these corresponded to the times at which clocks stopped, as earthquake shaking of a certain amplitude will often cause pendulum clocks to stop (the very property of early clocks that rendered them useless at sea!). But were these clocks stopped by the very first P wave, or by later waves? Dutton recognized the importance of this question, and addressed it at length. He further recognized the associated problems from both inaccurate clocks and imprecise recordings. From the long list of accounts, he singled out a mere half-dozen that he considered to represent the most reliable values. These include the report of an individual who was standing in front of an accurate jeweler's clock in New York City, waiting to set his watch, when he felt vibrations that he recognized to be an earthquake. He noted the time on the jeweler's clock, and even estimated the short delay between the time that he first felt the vibrations and noted the clock time. Not close to the 1/100 second accuracy that modern seismologists have come to expect, but not too shabby for 1886.

From Dutton's half-dozen most reliable observations, he estimated a speed of 5,205 meters per second. He obtained a very similar result, 5,192 meters per second, when he considered accounts that were very good but given only to the nearest half-minute. Not surprisingly, when he estimated a speed from the times at which clocks stopped throughout the country, the value was lower: 4,245 meters per second, consistent with the expectation that, at distant locales especially, clocks would not be stopped by the very first vibrations.

Seismologists now know that, observed across the surface of the earthquake, earthquake speed is not constant with distance from the epicenter. The speed of waves in rock increases with depth in the earth; the deeper one goes in the earth, the stronger the rock, and therefore the higher the speed of earthquake waves. Moreover, as one moves farther from an epicenter, the P and S waves that arrive at a site have traveled progressively deeper into the earth. The net result is that, when plotted against distance, the timing of both P and S waves does not show the simple linear slope that would be expected if the speed were constant. A secondary but sometimes important effect is that the speed of waves varies laterally in the earth as well, depending on the nature of the rocks through which the waves travel.

Nonetheless, one can refer to the seismological literature for a set of travel timetables: expected transit times of both P and S waves at different distances. Here one finds that an S wave at 10 degrees distance (1,110 kilometers) from an epicenter will have a travel time of 4 minutes and 17 seconds, or an average speed of 4,320 meters per second—differing by a mere 2 percent or so from the "inferior" estimate that Dutton obtained from his stopped-clock observations!

Interestingly, Dutton's "reliable" estimate of 5,200 meters per second differs more substantially from the average P wave velocity according to Richter's tables. According to modern estimates, a P wave recorded at a distance of 1,000 kilometers will have an average speed of about 7,500 meters per second.

What went wrong? One can only speculate, but almost all of Dutton's most reliable observations were from the Northeast; all six values were at distances of over 640 kilometers from Charleston. Had Dutton's estimate of the origin time of the shock at Charleston been too early, this would have skewed the estimate, but several reliable observations support Dutton's preferred estimate of the origin of the shock. More plausibly, it appears that observers in the Northeast did not always feel the initial P wave.

Notwithstanding either these limitations or Dutton's strong inclination to respect the limits of the data and his own understanding, he concludes chapter 7 satisfied that the work represents a solid contribution to the body of knowledge about earthquake waves. He notes the recency of the adoption of standardized time, and that such efforts in other countries have generally lagged those in the United States. He further notes the abundance of observations available for the Charleston earthquake compared with other large events, concluding that he has "no hesitation in expressing the belief that all [other estimates] that have ever been made and published hitherto possess much less weight than the data obtained from the Charleston earthquake."[14]

Dutton was aware that his result was considerably higher than earlier estimates obtained by Mallet and Milne, but pointed out that the earlier studies relied on data from "artificial tremors"[15]—vibrations generated at the surface of the earth and recorded at close distance. (Mallet attempted to measure the speed of waves from the 1833 Nepal earthquake but was stymied by a paucity of reliable clocks.) As discussed in the following section, Dutton understood that earthquake waves are elastic waves, and reasoned that, first, their speed should increase as rock density increases and, second, that rock density should increase with increasing depth in the earth. He therefore concluded that "It is not to be expected that the superficial layers of the earth will transmit the waves with so high a speed as the deeper layers; for their elasticity must be very much less."[16]

Even though Dutton's estimate reflected a measure of bias, and even though he did not understand every aspect of the problem with perfect clarity, he collected and analyzed the data with the utmost care, and made the right conclusion for the right reasons. This is what we call good science.

Having analyzed the timing observations, Dutton went on in chapter 8 to discuss other observations in light of newly developed theories of wave motion. In so doing, he not only brought theories to the attention of the American community, but also expressed them with singular clarity. The first part of the chapter should be required reading for every seismology course that tackles wave propagation from a modern—which is to say highly mathematical—point of view. Whereas such treatments often begin and end with equations, Dutton's discourse begins with, and rarely strays far from, the first principles that form the backbone of a theory.

Dutton begins by describing waves as "impulses" that proceed, or travel, from a disturbance. He goes on to discuss the familiar example of water waves—ripples generated in a body of water—which propagate as the initial disturbance affects the water in all directions from the point of origin. Such waves, he notes, are considered gravity waves, not in the more esoteric sense sometimes used in modern physics (waves that transmit the force of gravity), but simply in the sense that, once disturbed by a force, water returns to its original position because of gravity.

A moment of reflection reveals that all waves must represent an interplay between forces, because otherwise there would be nothing to counter an initial disturbing force, and therefore nothing to create the oscillation that defines the essence of a wave. Something pushes, something else pushes back; and the second something is not always gravity. A fundamentally different type of wave—Dutton first uses the example of a sound wave in air—is one in which the initial disturbance is countered by the elasticity of a medium. When scientists use the term "elasticity," they essentially mean the ability of a material to bend and then bounce back into shape. Push any material, and it will bend to some extent, but it will tend to snap back to its original form. A Slinky serves as a useful example: if one holds the toy taut and plucks the coil, the initial disturbance travels along the coil as each point disturbs the neighboring points, but once the disturbance has passed, the coil reverts to its former shape.

To understand earthquake waves, it is necessary to explore the concept of elasticity further, as Dutton proceeds to do. Among the concepts that he goes on to discuss is the fact that a material will be characterized by two different types of elasticity, the degree to which it snaps back when (1) its entire volume is subject to compression (imagine squeezing a sponge) and (2) its shape but not its volume is deformed. These two types of elasticity suggest, as Dutton reasons, that solid bodies are capable of transmitting two types of elastic waves. He terms these waves *normal* and *transverse*, the former resulting from elasticity of volume and the latter from elasticity of shape. Both of these terms reflect the motion of any given particle in a solid through which a wave passes: a normal wave will cause particles to move in or opposite to the direction that the wave is traveling, while a transverse wave will cause particles to move at 90-degree angles from the direction of wave propagation. Today seismologists refer to these as P waves and S waves, respectively. The identification of these waves, and their link to compression and shear forces, was left to Richard Oldham in 1897. In 1901 Oldham further noted the difference in the speeds of the two types of waves.

Again drawing from first principles, Dutton writes (quite correctly) that complicated mathematics is needed to fully describe the propagation of waves in solids. He refers to published theoretical studies, including the seminal work of Lame on acoustic (sound) wave propagation. But his elucidations, discussed here only in brief summary, capture much of the essence of wave propagation theory. In his discussion of the application of general wave theories to the particular case of seismic waves in the earth, Dutton's scientific acumen manifests itself as nothing short of prescient in anticipating the nature of these waves. (It is difficult, however, to separate original contributions from previously established ones.) Dutton writes that materials in the earth may not be perfectly *isotropic*, a word that refers to a constancy of properties with varying directions. Seismologists later identified evidence for anisotropic wave propagation—a full century, give or take, after Dutton's words were penned.

Dutton observes that no material is perfectly elastic, and deduces that the imperfect elasticity of rocks will cause a diminution of wave amplitude. Today we know this effect as *attenuation*. He talks about the importance of the earth's surface—in mathematical parlance a free surface—and describes how a different type of wave, the surface wave, will be generated. He notes that such waves would also be elastic waves, and therefore not the same as the waves that travel along the surface of the water. He further speculates, however, that actual gravity waves might be generated in shallow near-surface sediment layers, and notes that such waves would explain the common observation of high-amplitude waves rolling visibly across the ground during large earthquakes. This supposition has been echoed in modern-day seismological discourse; to this day, however, it remains unclear if and under what circumstances such waves might be generated. Harry Fielding Reid argued that such visible undulations could in fact be elastic waves traveling in weak sediment layers near the surface.

In other respects Dutton's musings on the effect of near-surface sediments are directly on the mark, as defined by modern theories. Recognizing that shallow sediments have lower elasticity than more compact, solid rock at greater depths, Dutton concludes that earthquake waves will decrease in speed and increase in amplitude once they travel into shallow sediments. He further points out that the period of the waves will be affected as well, describing the essence of a phenomenon that we know today as resonance.

Through most of his final chapter Dutton concentrates on an overview of seismology theories established at the time. In his final pages he uses his estimate of wave speed to estimate the wavelengths of earthquake waves as roughly 519–1,037 meters. Because his estimate of wave speed was higher than previous estimates derived from near-surface experiments, Dutton was the first to appreciate the large wavelength of earthquake waves.

Dutton's final chapter concludes with speculations on the quantification of earthquake waves. He describes two quantities of potential utility to measure waves: acceleration, and the energy per unit area of a wave front. He observes, once again quite correctly, that the latter quantity will be a complex function of several factors: the amplitude, wavelength, and period of a wave, and the density of the medium in which it travels. But he concludes, "To me it seems unquestionable that [energy per unit area] is to be preferred."[17] Dutton's remarkable report thus ends, appropriately, on another note of remarkable prescience. While the seismological and engineering communities relied on ground acceleration as a preferred metric for ground motions through much of the 20th century, they have increasingly recognized the limitations of this approach, and the necessity of developing energy-based methods with which to quantify shaking.

Dutton's final summary is not at the end of his report, but in the final paragraph of its preface. Echoing Samuel Mitchell in both content and eloquence, Dutton wrote:

But after the most careful and prolonged study of the data at hand, nothing has been disclosed which seems to bring us any nearer to the precise nature of the forces which generated the disturbance. Severe labor has been expended for many months in the endeavor to extract from them some indications respecting this question, but in vain. This problem remains where it was before. Having nothing to contribute towards its solution, I have carefully refrained from all discussion of speculations regarding the causes of earthquakes.[18]

While the modern reader cannot help but admire the integrity behind these words, the modern scientist cannot help but feel a pang of regret over the speculations that Dutton clearly entertained but chose not to voice. One imagines they would have made for good reading.

Faults in Charleston

The Charleston region has been a high priority for earthquake hazard research in the United States, but to a lesser extent than the New Madrid region, north of the present-day city of Memphis. This disparity has to do with the exigencies of both politics and science. Although now considered fairly close in magnitude, the largest New Madrid earthquakes were once thought to be quite a bit larger than the Charleston earthquake, and the New Madrid seismic zone was thus believed to represent a greater potential hazard. But initial modern fault-finding investigations of the New Madrid seismic zone also proved more fruitful than those in Charleston, owing to a number of factors, not the least of which is the success of the earliest fault-finding investigations. In scientific research, success begets success. When Jack Odum and his colleagues concluded that the pattern of disruption along the Mississippi River argued for thrust motion on the Reelfoot fault, this gave other earth scientists a target at which to aim investigations using other methods. When Arch Johnston and his colleagues pieced together a detailed rupture scenario for the 1811–1812 sequence, this provided the grist for testing theories of fault interactions. And so forth. Before a line of scientific inquiry can pick up steam, scientists need to establish a beachhead: initial results that lead to testable hypotheses, which often lead in turn to further results.

In the early 1970s the U.S. Geological Survey published as a Professional Paper a collection of research papers on the Charleston region and the 1886 earthquakes. This volume includes a paper by Gil Bollinger that presents a thorough analysis of the shaking effects of the temblor throughout both South Carolina and the country as a whole. Most of the papers in the volume focus on a general geologic and geophysical assessment of the Charleston region. These studies include such characterizations as the magnetic, geologic, and chemical properties of the crust, characterizations that are critical to understanding the structural framework of a region—even though the report has relatively little to say about the fault(s) that had ruptured during the 1886 temblor or the details of the 1886 event itself. Which is to say that before scientists can find and understand faults (especially in a complicated region), it is first necessary to understand the overall lay of the land.

Charleston sits atop the Coastal Plains sediments, a wedge of sedimentary rocks that begins at the eastern edge of the Appalachian Mountains and ends at the coast. These sediments, deposited over millions of years, reach a thickness of over a kilometer at the coast. By the early 1970s, scientists had determined the nature of the Coastal Plains sediments and deduced that the 1886 earthquake likely occurred on a fault, or faults, in the crystalline basement rock below the sediments. Such a setting tends to frustrate scientists' efforts to find faults, obviously, because surface observations will be limited and, as a rule, ambiguous. As was largely the case at New Madrid, the surface disruptions so faithfully documented by Earle Sloan were only secondary features of a fault rupture that did not reach the earth's surface.

Reading through the USGS Professional Paper, one learns a lot about the overall geologic setting of the Charleston region, but markedly little about the fault(s) that produced the 1886 earthquake. Few hints of fault structure were found by any of the studies. One study of recent small earthquakes in the region did yield a number of clues, however. A small network of seismometers was deployed and operated between 1973 and 1975; these instruments recorded 13 small temblors, most too small to be felt but all large enough to be located using standard analysis procedures. When plotted on a map, the epicenters of many of these events were aligned on a northwest–southeast axis, the northwest terminus of which was near Summerville.

The largest earthquake recorded by the seismometers, a magnitude ~4 earthquake on November 22, 1974, was analyzed using more sophisticated (for the time, relatively new) techniques to obtain a so-called focal mechanism. Focal mechanisms are determined from observations of P wave polarity: simply whether the very first P wave recorded on a seismometer is up or down. Any given fault rupture, or earthquake, will generate a characteristic pattern of ups and downs; observed patterns can thus be used to determine the orientation of an earthquake rupture. One complication, however, arises from the nature of earthquake ruptures and wave propagation: focal mechanisms always yield two possible rupture orientations; it can be very difficult to choose between them. In the case of the 1974 shock, however, one of the

two orientations—a nearly vertical plane oriented northwest–southeast—
was preferred because it was more consistent with the locations of small aftershocks and because the strongest shaking occurred over an ellipse with a
similar orientation.

Scientists were thus able to identify with a measure of confidence the
approximate location and orientation of a small earthquake that occurred
near Charleston in 1974, and to conclude that the temblor occurred on a
northwest-trending fault between Charleston and Summerville. To understand the limitation of this result for the larger question of understanding
the 1886 earthquake, consider just how much—or, more precisely, how
little—we would learn by inferring the fault parameters of a stray M4 earthquake that occurred somewhere in the vicinity of the San Andreas fault,
nearly a century after the 1906 earthquake. To answer this question, a review
of a few earthquake ABCs might first be in order. Because the size of an
earthquake reflects the area over which a fault moves, and the rupture area
of an M4 earthquake is usually less than 1 kilometer across—not a great distance, considering that such earthquakes almost always occur at least 5 kilometers deep. And whereas very big earthquakes cannot occur without very
big faults, the crust is literally riddled with small faults, almost all of which
are capable of generating earthquakes of magnitude 4. In California, it is perhaps the rule rather than the exception that small earthquakes that occur
close to large faults are actually on much smaller, secondary faults. So by
studying a stray small event, even in great detail, we may well learn almost
nothing about faults that generate large earthquakes in the same region.

Moreover, in many areas, the faults that generate large earthquakes do not
seem to generate many small ones. Along segments of the San Andreas fault
known to rupture in very large earthquakes, scientists today observe extremely few small earthquakes that are actually on (as opposed to adjacent
to) the fault. At New Madrid there is reason to believe that small earthquakes
are indeed occurring on the faults that ruptured in 1811–1812, but is it reasonable to expect that the major faults near Charleston are following this model
rather than that of the San Andreas? The assumption is not grossly unreasonable. In geologic terms South Carolina has more in common with New Madrid
than with California, since New Madrid and South Carolina are both well
away from the nearest active plate boundary. According to current theories,
where aftershocks and faults are concerned, this commonality trumps the
fact that South Carolina and California are both coastal locations.

Combining the above ambiguities with the basic fact that small earthquakes occur far less commonly in South Carolina in than New Madrid, one is left with a disheartening prognosis concerning the use of small modern earthquakes to investigate large faults in the region. That is, it is not at all clear that the smattering of small earthquakes in the Charleston region can tell us anything about the large earthquake that struck over a century ago.

The nature of earthquake rates in processes in regions such as South Carolina is, however, such that scientists cannot afford to be too easily discouraged. In South Carolina, such perseverance has indeed borne fruit, illuminating at least some faults in the area. The Ashley River fault system, which includes the Ashley River fault as well as the Sawmill Branch fault, extends on a northwest–southeast axis along the Ashley River for some 30 kilometers roughly between Summerville and the town of Magnolia Gardens. The river itself is very likely there because the fault is there or, rather, both are there for the same reason. Very typically, zones of weakness in a midcontinent region are associated with both major river valleys and faults. Small event locations and other observations have illuminated a second active fault as well, the Woodstock fault, which may have two active strands, one to the north–northeast and one to the south–southwest of the Ashley River fault. Results suggest the Woodstock fault is as much as 100 kilometers long.

The Woodstock and Ashley River fault systems represent a pair of intersecting features, leading seismologist Pradeep Talwani to propose that intraplate earthquakes will preferentially occur in regions where such pairs of intersecting faults are found. According to this theory, fault junctures represent not only points of mechanical weakness in the crust, but also areas where rocks are generally more fractured, which in turn implies greater rates of fluid flow in the crust. In the midst of a broad and strong region of crust, fault intersections may therefore become the loci for concentrations of stress, giving rise to heightened rates of earthquake activity.

Not unlike the theory that stresses are concentrated at New Madrid because of the shape of subsurface layers in the crust, Talwani's theory has a measure of intuitive, geometric appeal. Certainly the inferred faults succeed in explaining the distribution of effects, and inferred "epicentrum" locations, identified by Earle Sloan. The primary epicentrum, according to Sloan's account, falls nearly along the Ashley River fault, some 30 kilometers northwest of the town. As noted previously, judging from the abrupt, booming nature of the foreshocks in Summerville, the northwest end of this fault has to be

FIGURE 6.6. Wrecked house in the village of Summerville, South Carolina. Although wood-frame houses tend to fare well in even strong earthquakes, many such houses in the area were built with raised foundations, which leaves them vulnerable to toppling. (Figure 9 from Clarence Edward Dutton, "The Charleston Earthquake of August 31, 1886." *Ninth Annual Report of the United States Geological Survey, 1887–88.* Washington, D.C.: Government Printing Office, 1889)

the prime suspect for the epicenter (as defined today) of the 1886 main shock (figure 6.6). That is, classic foreshock sequences, which involve sequences of small earthquakes tightly spaced in both space and time, are commonly—if not always—observed to occur in proximity to the initiation point of the earthquake. Seismologists do not yet understand the processes that are at work when a major fault approaches failure. One theory says that small earthquakes, or foreshocks, will simply cascade into bigger earthquakes, just as adjacent dominoes will topple each other. A second, more interesting theory says that faults will start to slip very slowly and gradually, triggering a smattering of small earthquakes before detonating the main event.

Several factors suggest that the foreshocks occurred extremely close to Summerville. First we have the accounts of the booms that initially impressed many residents far more than ground shaking. Recall that earthquakes generate shaking at a range of frequencies, including high frequencies—akin to the high tones in music—that edge into the audible range for humans. Such high-frequency energy is quickly damped out as it travels in the crust; earth-

quakes are not generally audible (apart from secondary shaking effects) unless they are extremely large or small and extremely close to an observer. The
shaking from small earthquakes, meanwhile, tends to be fairly simple at close
distances—often a single *whump* generated by a short burst of motion on a
fault. As one moves away from the epicenter of a small earthquake, the overall amplitude of shaking diminishes (recall the foreshocks were barely felt in
Charleston), but the nature of the shaking often becomes more complex and
prolonged as waves are scattered around in the crust.

Building on the earlier discussion, one can thus summarize a rupture scenario as follows. On or about August 27, 1886, the northwest edge of the Ashley River fault commenced a countdown sequence. Whether small earthquakes were triggered by an underlying process or simply nudged one another
along, the sequence eventually cascaded into catastrophic failure: substantial
rupture on the Ashley River fault, originating near the village of Summerville and, following a few seconds of gradual acceleration, careening at a high
rate (about 3 kilometers per second, or 100 miles per minute) toward the city
of Charleston. As is typically the case with large earthquakes, the rupture
picked up steam as it went along, not speeding up but causing increased motion, or slip, along the fault, which in turn gave rise to increased shaking. The
"epicentrum" near the village of Woodstock appears to mark the point at
which the rupture hit its crescendo before continuing on, with somewhat less
oomph, and petering out somewhere northwest of the city of Charleston.

This sort of rupture would have had two immediate and drastic effects.
First, by virtue of the Doppler effect that causes trains to sound different
coming versus going, the direction of the rupture to the southeast would
have resulted in especially strong shaking in Charleston. Seismologists call
this a directivity effect. As far as directivity effects are concerned, the city of
Charleston could scarcely have been in a worse location.

The second major effect of an Ashley River fault rupture is somewhat
more complicated, and requires another bit of earthquake ABCs. Every large
earthquake has a substantial effect on the surrounding crust by virtue of the
fact that large parcels of real estate are being rearranged. Since the early 1990s,
an exciting subfield of seismology—earthquake interactions—has involved
the detailed investigation of this phenomenon, focusing on the pattern of
stress that is left behind when a large earthquake occurs. This stress redistribution appears to account for the spatial distributions of aftershocks and,
sometimes, the spatial and temporal distribution of subsequent main shocks.

Moreover, seismologists now recognize a complicated gray area between the simple earthquake taxonomy implied by terms such as "aftershocks" and "main shocks." For example, if a large aftershock occurs within seconds, or fractions of a second, of a main shock, the aftershock is considered to be a *subevent* of the main shock, even if the two occur on different faults. The M6.5 1992 Big Bear earthquake, itself an aftershock of the M7.3 Landers, California, earthquake, appears to have been one such composite event. According to the recently developed theories of earthquake interactions, subevents are easily understood as, essentially, large aftershocks that are triggered almost immediately by an initial main shock rupture. And when such composite earthquakes do involve two markedly different faults, they are not uncommonly in the same geometry as is found in Charleston: at nearly right angles to one another. This phenomenon is thought to result from the stress redistribution caused by the initial rupture, a redistribution that not only can push some neighboring faults toward failure but also can essentially unclamp other faults, thereby releasing the stress that locks them together and allowing them to generate their own earthquakes. This unclamping tends to work when two faults are at right angles close to one another. As modern seismologists have developed sophisticated methods to determine fault motion, we find a perhaps surprising number of earthquake ruptures that resemble Ts or Ls. Such complex events are not considered to be two separate earthquakes (even though two separate faults as involved) as long as the time delay between the two fault ruptures is a few seconds. (In 1987, earthquakes struck on two perpendicular faults in southern California, with a time lag of about 12 *hours*. Nobody suggested these should be considered one earthquake, but the difference between these events and the Charleston earthquake was probably only a matter of degree. Which is to say that sometimes semantics looms large in the field of seismology.)

Considering the inferred geometries of the Woodstock and Ashley River faults, one can then piece together the second half of the main shock rupture scenario. The lateral rupture on the Ashley fault would have unclamped the Woodstock fault, triggering a rupture that most likely started near the intersection of the two faults. This rupture then picked up steam as it sped to the south (and possibly north–northwest as well)—mercifully away from Charleston. (Two segments of the Woodstock fault have been identified.)

Considering the lengths of the faults and the mapped shaking effects, it appears that the Ashley River fault ruptured over a distance of about 33 kilometers. The Woodstock fault rupture may have been about the same size, although we have little direct information to determine its length. Further considering the speed at which earthquake ruptures are known to propagate along faults, each of the two faults would have been in motion for no more than 15 seconds. The actual movement of a fault is virtually always much faster than the duration of strong shaking, a consequence of the reverberation of waves once they leave the fault. Still, comparing this against the intensity and duration curve painstakingly gleaned from the observations by Earle Sloan, which includes two pronounced peaks of strong shaking, one can conjecture that the two ruptures were separated by a few, no more than about ten, seconds. A similar delay (about four seconds) was estimated for the subevents of the Big Bear earthquake.

How big was the Charleston earthquake? This question is complicated by the realization that two separate faults were involved: the energy release of such a multiple event is different from that of a single earthquake with the same combined length. The question is further complicated by the fact that magnitude does not correlate perfectly with rupture length, but also depends on the amount that a fault slips. Earthquakes with a given rupture size can have marked differences in slip, which is to say that some earthquakes pack somewhat more of a wallop than befits their size. Moreover—with intraplate earthquakes there always seems to be a "moreover"—it appears that intraplate earthquakes tend to pack more of a wallop than their similar-sized brethren in active plate boundary zones.

Based on typical results from California, an earthquake with a rupture length of 33 kilometers will have a magnitude of perhaps 6.8–7; combining two such ruptures doubles the energy, which implies a +0.3 increase in magnitude by virtue of the logarithmic nature of the magnitude scale, or an overall magnitude of 7.1–7.3.

How does this compare with modern estimates? Awfully well. In contrast to the New Madrid earthquakes, the magnitude of the Charleston earthquake has not been the subject of inordinate controversy over the years. Given Gil Bollinger's detailed analysis of intensities and several modern methods to calibrate the earthquake with data from more recent temblors, several published studies have inferred magnitudes in the same range that one obtains from the simple, back-of-the-envelope calculations above.

Thus it appears that, despite the relative paucity of scientific investigations aimed at the 1886 earthquake, scientists have made substantial strides in their quest to understand faults in and around Charleston. By investigating prehistoric liquefaction features (see sidebar 6.1), scientists have identified prehistoric earthquakes of apparently the same size as 1886. The two best-

SIDEBAR 6.1

In California, geologists are able to dig into faults just below the earth's surface and, using the pattern of disrupted sediments, piece together a chronology of prehistoric earthquakes. In both Charleston and New Madrid, faults either don't reach the surface or don't reach it in any simple way. To conduct so-called paleoseismology studies in these regions, geologists instead search for liquefaction features such as sand blows and sand dikes to estimate dates of prehistoric earthquakes, and sometimes to learn more about temblors that struck during historic times. In the Charleston region, Pradeep Talwani and his colleagues have found evidence that large earthquakes—comparable in size to the one in 1886—strike about once every 400–500 years.

documented prehistoric events occurred around A.D. 1500 and A.D. 1000. Curiously, all three of these dates follow by 50–100 years the known or estimated dates of large New Madrid sequences, which occurred around 900, 1450, and 1811–1812.

The apparent coincidence of event dates at Charleston and New Madrid raises a most intriguing question: Are these two fault systems, separated by over 1,000 kilometers, somehow in communication with one another? Not too long ago scientists would have dismissed the notion out of hand, because even large earthquakes were believed to produce no effects of any consequence at such great distances. Now we cannot be so sure. The New Madrid earthquakes triggered separate earthquakes at distances of at least 500 kilometers, in northern Kentucky, and some—albeit ambiguous—evidence suggests they may have triggered small earthquakes in Charleston as well. Scientists understand how earthquakes are triggered within seconds to a few days following large main shocks. Could such triggering initiate a process that culminates in large earthquakes, but after a delay of several decades? Perhaps, but we do not yet understand how.

The study of earthquake interactions is now in its infancy; many fundamental questions remain unanswered. If finding faults can be a challenging exercise, understanding them—not only their structure but also their dynamics and interactions—is vastly more difficult still. Concern for earthquake hazard might provide the impetus for such research, but such are the intriguing and vexing questions of which exciting science is made. Scientists' quest to understand important historic earthquakes contributes to our quantification of future earthquake hazard, but not only that. Carefully gleaned data from some of the oldest known earthquakes can provide fundamental insights into some of our newest science. Our planet is a fantastically complicated place; even our most sophisticated theories invariably represent a gross simplification of processes at play in the real earth. Given this, scientists would be loath to propose, based only on theory, a notion as outlandish as the communication of fault zones separated by 1,000 kilometers. It is up to the earth itself to reveal its best surprises. It is therefore the obligation of scientists to pay attention, not only to the data of today but also to the stories of the past, and to the remarkable storytellers who wrote them.

Following the 1811–1812 New Madrid earthquakes, the best storytellers were a small handful of educated individuals who had little or no background in earth sciences of any stripe; at that time there was no such thing as a background in seismology. By 1886 the landscape had changed in many ways, not the least of which (at least in this context) was the fact that the earthquake was investigated and chronicled by individuals who did know some things about earthquakes. The earliest seismological research focused on the waves generated by earthquakes, and so the focus of the Dutton report was shaped fundamentally by its context. Dutton's investigations tested prior theories and results, such as the speed of earthquake waves; his interpretations and speculations, which reflected both acumen and vision, also helped propel the field forward.

In the end Dutton found himself as frustrated about the causes, and indeed the basic mechanics, of earthquakes as Samuel Mitchell had been 70 years earlier. It could scarcely have been otherwise, given not only the state of understanding of faults and earthquakes but also the absence of dramatic surface faulting that might easily have led a man of Dutton's intellect to develop theories that did not appear on the scene for another two decades. Like any science, the field of seismology advances collectively. Rapid strides in understanding, even revolutions, occur when a basic framework is in place and, typically, when new data suddenly cast illumination where before there had been darkness, or perhaps the murky haze of unproven ideas. This is true for any scientific field; the wrinkle in earthquake science is that the earth itself produces many of the most critical data, and on its own time.

Every historic earthquake is unique in its own way, and not only in the ways that come obviously to mind. Many of the important historic earthquakes of the 19th and early 20th centuries played critical roles in the development of earthquake science as a modern field of inquiry. The nature of the roles depended on the vagaries of the earthquakes themselves, as well as on the inclinations and talents of the remarkable individuals whose attention they captured.

7

Finding Faults in California

All combines in one composite impression,—hurry and
hopefulness, gaiety, silken petticoats and starched gowns,
corduroys, tramping boats, sombreros, temporary wooden
buildings, the flutter of many flags, rush of automobiles, clatter
of lumber, banging of hammers and the rumble of drays,—the
very sunlit air seems to breathe renaissance. This is the spirit at
the Golden Gate,—the old spirit that still lives—and this is why
it is fun to live in the new San Francisco.
> —Col. Edwin Emerson, "San Francisco at Play,"
> *Sunset* magazine (October 1906)

April 18, 1906. "At 5:15 this morning . . . I thought I heard the alarm go off. I
reached over to stop it and to my great surprise it was rolling from one side
of the stand to the other, & then to the floor. I looked out the window . . . in
time to see a few chemnies [*sic*] sway around and fall. The picture & bed &
dresser & chairs were dancing around the room. . . . A house caught fire
about 5 blocks off. . . . Then to make matters worse, there was no water when
the fire dept. arrived."[1]

April 18. "Within moments, during this period of the city's greatest emergency, the unusual silence of the [fire] alarm bell told its own story. The system was destroyed as was the function of the city's 30,000 telephones."[2]

April 18, 7:00 A.M. "The Federal Troops, the members of the Regular Police Force and all Special Police Officers have been authorized [by San Francisco Mayor E. E. Schmitz] to KILL any and all persons found engaged in Looting or in the Commission of Any Other Crime."[3]

April 19. "I have seen the most awful sights to day that I ever saw in my life! . . . It is impossible for you to conceive or in any small degree realize the terrible disaster that has befallen San Francisco. I can't & I've seen it. . . . When I left this afternoon fully 2/3 of San Francisco was in ruins. The streets have great cracks in them & the Car tracks are twisted by the earthquake & heat. The flames are spreading in all directions even against a fresh north wind."[4]

May 5, 1906. "Day and night the dead calm continued, and yet, near to the flames, the wind was often a gale, so mighty was the suck."[5]

May 13. "We have not got our thoughts collected since the big quake—not quite—it has been 24 days since the big awful earthquake and we have had more then 24 earthquakes in them 24 days, small ones. We can only hope that we have seen the worst of it, but we can't help looking for another big one anyhow. . . . I wouldn't come to this state until it got so that it would hold still."[6]

May 1906. "A liveryman who drove me about the city tried to explain this jubilant acceptance of misfortune. 'San Francisco was the gayest city in the world,' he said, 'and it's hard for these people to take trouble seriously.'"[7]

In retrospect one cannot help but be struck by the apparent dichotomy between the horrors of eyewitness accounts of devastating earthquakes and the optimism that shines through words penned only days later. In a book focused largely on the latter, the reality of the former must still be acknowledged. Earthquakes are powerful events, easily one the most terrifying natural disasters faced by mankind. Earthquake-savvy Californians may grow complacent about small tremors, but when the Big Ones, or even the more common Pretty Big Ones, strike, the reaction is as universal as it is primal: terror. When the dust settles, the immediate aftermath of an earthquake in an urbanized society can be profound. A century after 1906 and a quarter of a millennium after 1755, earthquakes still have the ability to shake, and scare, the living daylights out of people.

Countless eyewitness accounts tell the story of the 1906 earthquake, which struck without warning at 5:12 A.M. on April 18. The temblor itself is generally described as a minute of horror, one that sent furniture crashing, left buildings twisted and toppled, and tore the earth apart along the fault and elsewhere that soft soils buckled and gave way. As destructive as the shaking was, the earthquake itself left San Francisco and other cities along the fault bruised but not beaten. Had the story ended when the shaking stopped, the legacy of 1906 would be far different from the one we know today. But as accounts tell us, the silence that followed the shaking was deafening and ominous: the city and its water supply were too badly battered to mount an effective response. Citizens could do little more than stand back and watch the city burn nearly to the ground. From Van Ness Street eastward, very little remained standing by the time the flames ran their course three days later.

The death toll from the 1906 temblor was surprisingly low. A 1972 NOAA report, based largely on a 1906 report of U.S. Army relief operations, estimated a direct death toll of 700–800, 498 of the dead in the city of San Francisco. Following an exhaustive search of public records, Gladys Hansen, archivist for the City and County of San Francisco, estimated a significantly higher number: closer to 3,000. Hansen's tally includes deaths that could be attributed to the earthquake according to standard criteria for assessing fatalities following natural disasters, including deaths from suicides attributed to depression and from certain diseases (in this case, including smallpox). Still, even this higher figure is comfortably below the number of Americans who now lose their lives in automobile accidents in an average month.

In *Denial of Disaster*, Hansen and Emmett Condon conclude that the impact was deliberately and substantially underplayed in the aftermath of the earthquake by those with a vested interest in the city's recovery. The San Francisco Real Estate Board passed a resolution that "the great earthquake" should be thereafter referred to instead as "the great fire."[8] Hansen and Condon also describe the letter-writing campaign of James Horsburgh, Jr., of the Southern Pacific Company, who urged chambers of commerce throughout the state to portray the disaster as primarily the consequence of a large fire, and to emphasize the city's rapid and energetic recovery. The company went on to publish the glossy *Sunset* magazine (no relation to the modern publication of the same name), which, in Hansen and Condon's words, provided a "sanitized, simplistic, and, in many cases, grossly inaccurate version of the earthquake's effects."[9] This slickly packaged material found its way into subsequent publications.

Scientists and engineers also found themselves pressured. John Caspar Branner was a Stanford professor of geology and one of the driving forces behind the creation of the Seismological Society of America in the months following the earthquake. In a 1913 publication in the Society's *Bulletin*, Branner recounted his and his colleagues' experiences following the earthquake:

Shortly after the earthquake of April 1906 there was a general disposition that almost amounted to concerted action for the purpose of suppressing all mention of that catastrophe. When efforts were made by a few geologists to interest people and enterprises in the collection of information in regard to it, we were advised over and over again to gather no such information, and above all not to publish it. "Forget it," "the less said, the sooner mended," and "there hasn't been any earthquake" were the sentiments we heard on all sides.[10]

Clearly Branner and his colleagues knew from whence they spoke. Business leaders of the day—and therefore political leaders of the day—had a strong, vested interest in downplaying not only the severity of the disaster but also California's future earthquake risk. "The idea back of this false position—for it is a false one—," Branner wrote, "is that earthquakes are detrimental to the good repute of the west coast, and that they are likely to keep away business and capital, and therefore the less said about them the better."[11]

Page after page of photographs in *Denial of Disaster* tells the true story of the disaster: a city reduced to rubble and ash, devastation the likes of which "the world would not see again until the bombing of Hiroshima in 1945"[12] (figure 7.1). A photograph taken from a balloon six weeks after the temblor reveals a ghostly scene: the ruins of City Hall to the west, tall buildings in various states of disrepair in the central business district, and block after block of rubble everywhere else. By this time, many severely damaged structures had been dynamited into smithereens. The loss of life might have been modest, but the devastation was staggering.

Earthquakes take a terrible toll on cities and people: they can lay waste to lives, homes, and businesses. The nature of the toll—the human losses compared with the financial ones—differs sharply in different parts of the world. Earthquakes such as the 1906 and the 1994 Northridge claim relatively few lives, but cause exorbitantly expensive property damage. In other parts of the world the proportions can be reversed. Where construction materials such as mud brick and stone are abundant and cheap, it may not be expensive to rebuild houses after earthquakes. In many cases survivors simply restack the same stones on top of each other. In such regions it is the toll of human life that can be staggering.

But is the story of San Francisco's recovery truly, and only, one of denial? Of a prettied-up face presented to the world by a railroad company hell-bent

FIGURE 7.1. Ruins of San Francisco in the aftermath of the earthquake and fire. View looking west from Telegraph Hill. (Photograph by Frank Soule; courtesy of U.S. Geological Survey)

on protecting its own interests? While some in California may have sought to minimize the impact of the earthquake by itself, the fact remains that the fires, not the earthquake, caused the lion's share (90–95 percent, by most estimates) of the property damage. The damage from shaking was also highly concentrated along the fault line. In the words of pioneering geologist Grove Karl Gilbert, "At a distance of twenty miles [from the fault] only an occasional chimney was overturned, the walls of some brick buildings were cracked, and wooden buildings escaped without injury; the ground was not cracked, landslides were rare, and not all sleepers were awakened. At seventy-five miles the shock was observed by nearly all persons awake at the time, but there were no destructive effects."[13]

For residents of San Francisco, the cause of the destruction mattered little; what mattered was the bottom line: much of the city lay in ruins. The story of San Francisco's rebound from this calamity is far deeper—and also far more heartening—than the part of the story that involves cover-up and shrewd marketing on the part of business interests. As French Strother went on to write (following his perhaps prescient "gayest city" quote):

> the explanation lay deeper. San Francisco had always been a generously hospitable city and an open-handed friend of others in distress. And intuitively San Franciscans knew that their misfortune instantly became a common misfortune to all the world. In fact, they suffered less, emotionally, than those who sympathized while they aided. In Fresno, two hundred miles away, and uninjured, strong practical men told me that when they saw the ruins they broke down and cried. "It was our city," was their explanation. Besides their own pluck, it was this sense of their community of interests with the rest of the country that restored instant confidence among the people of San Francisco.[14]

In San Francisco, as in Charleston in 1886 and Lisbon in 1755, people drew on the innately human traits of resiliency and resourcefulness to bounce back in the wake of a disaster of seemingly biblical proportions. Psychologist and philosopher William James, then a professor at Stanford, wrote of this resilience and compassion with singular insight and clarity: "The cheerfulness, or at any rate the steadfastness of the tone, was universal. Not a single whine or plaintive word did I hear from the hundreds of losers whom I spoke to."[15] James went on to say:

> It is easy to glorify this as something characteristically American or especially Californian. Californian education has, of course, made the thought of all possible recuperations easy. In an exhausted country, with no marginal resources, the outlook on the future would perhaps be darker. But I like to think that what I write of is a normal and universal peculiarity. In our drawing rooms and offices, we wonder how people ever *do* go through battles, sieges, and shipwrecks. We quiver and sicken in imagination and think those heroes superhuman. Physical pain, whether it be suffered alone or in company, is always more or less unnerving and intolerable. But mental pathos and anguish, I fancy, are always effects of distance. At the place of heavy action where all are more or less concerned together, healthy animal insensibility and heartiness take their place.[16]

James concluded, "At San Francisco, there will doubtless be a crop of nervous wrecks before the weeks and months are over; but meanwhile, the commonest men, simply because they *are* men, will keep on singly and collectively showing this admirable fortitude of temper."[17]

Pauline Jacobson, a staff writer for the *San Francisco Bulletin*, summed up the experience as "twenty-eight seconds of awful interrogation and then the most sociable time I've ever had in my life."[18] She went on to say, "I still stand by that summing up of that morning, a time long before things were at their worst, long before a red yellow-licking, leaping devil-flame had started on its mad revelry to add the greatest insult to the greatest injury that had yet been done a people."[19] (In this last sentence the reader does have to forgive a lack of awareness of, or appreciation for, history: a certain parochialism seeps out from around the edges of 1906 accounts from time to time.) "Most of us since then have run the whole gamut of human emotions from glad to sad and back again, but underneath it all a new note is struck, a quiet bubbling joy is felt. It is that note that makes all our loss worth the while. It is the note of a millennial good fellowship."[20]

Of course one should not look back through glasses of an overly rosy hue. The mayor's infamous "shoot to kill" proclamation was motivated by acts of looting witnessed by the mayor himself in the immediate aftermath of the temblor. The mayor's action still engenders debate, but almost surely discouraged further acts of lawlessness. Whether or not one is overwhelmed with a sense of millennial good fellowship, presumably one doesn't steal if stealing is likely to get a person shot on sight. Other survivors clearly did not fare well in the months and years following the earthquake. Among the list of temblors tallied in the Museum of the City of San Francisco's list "San Francisco Earthquake History 1880–1914" is the following item, dated nearly a year after the earthquake: "New York City police were on the lookout for San Francisco attorney Walter C. Stevens, who lost everything in the earthquake, became despondent, and was given money by the relief committee to go East. It was feared he would commit suicide."[21]

Nonetheless, in *Three Fearful Days*, a book of firsthand accounts compiled by Malcolm Barker, a remarkably consistent tone shines through. Tales of despondency are few and far between in these pages. The skeptic might observe that the truly despondent or suicidal, such as attorney Stevens, were not likely to have penned their remembrances. Still, "despondent" is not among the adjectives that one finds in many of the essays in Barker's compilation that describe the "man in the street" disposition. Instead, there are words of a different flavor: equanimity, humor, gaiety. Journalist Winifred Black Bonfils wrote, after describing the random acts of kindness that she encountered from total strangers:

> And then I knew that the dreadful story of death and hopeless misery the blackened ruins were trying to tell me was false. San Francisco, the best loved of the world, is not dead, and can never die while one man or woman with the true spirit that made the old San Francisco what it was still lives.
>
> The beautiful streets, the smiling parks, the friendly houses of friends, the gay restaurants—these things were only a little bit of the outside dress of San Francisco. The real San Francisco is just as much alive to-day as it was some seven sweet years ago when the whole city was gay with flags to welcome our boys home from the Philippines.
>
> San Francisco in ruins!

Why, you couldn't kill San Francisco with a dozen earthquakes and a hundred fires.[22]

In "The California Earthquake of 1906," Mary Austin contributed personal reflections to the compilation of chapters by noted scientists of the day. She wrote of the "well-bred community that poured itself out into Jefferson Square"—a community that never "forgot its manners" in spite of being "packed too close for most of the minor decencies." Austin added: "Right here, if you had time for it, you gripped the large, essential spirit of the West, the ability to dramatize its own activity and, while continuing in it, to stand off and be vastly entertained by it. In spite of individual heartsinkings, the San Franciscans . . . never lost the spirited sense of being audience to their own performance."[23]

In addition to this human and/or Californian capacity for resilience (and, one is tempted to add, stage drama), by the time the Big One struck San Francisco, the tradition of state-sponsored disaster recovery had long since been established. The U.S. Army made an enormous commitment of supplies and manpower to the recovery effort. On the day following the earthquake, Secretary of War (and future President) William Howard Taft directed 200,000 rations in response to a cabled appeal for tents and food. By April 20, the quartermaster general cabled that "all tentage of Army now en route to San Francisco. Contractors have been urged to hasten deliveries of duck."[24] By June 1906, some 10 percent of the standing Army was on duty in San Francisco to direct the relief effort.

The hastened duck and later relief efforts notwithstanding, San Francisco did not, of course, rebound overnight. Here again, it is clear that rebound comes at a cost. Tens of thousands left San Francisco following the earthquake, but 250,000 remained—homeless, dependent on relief efforts for survival. Those who could not find shelter elsewhere settled in tent cities as well as temporary lodgings in military facilities. The Army-run tent cities were relatively orderly and sanitary compared with the makeshift encampments that sprang up elsewhere throughout the city. Sanitary conditions deteriorated in the makeshift tent cities over the months following the earthquake. Although they did not reach epidemic proportions, early cases of plague, typhoid, and smallpox were followed in 1907 by an outbreak of spinal meningitis. Over time, some of the tents gave way to Spartan wooden cottages: small redwood shacks painted "park bench green," built at a cost of

$100 to $740 and rented for $2 per month. To lessen the chance that these settlements might threaten to become permanent, the city offered a rent-to-own program whereby a $50 sale price for each cottage was paid after two years, at which point the owners were required to move their new homes away from city parks. More than 5,000 cottages migrated to new sites under the program; by 1909 the last of the temporary settlements was closed. In the words of Rufus Steele, "The refugee shacks literally took wings, flew away to pleasant hillsides in the outlying districts, and there were transformed into comfortable houses."[25]

Those with the fewest resources and general wherewithal lingered longest in tents and cottages; elsewhere throughout the city, rebuilding efforts proceeded at a healthier clip. If business interests had a vested interest in playing up the elasticity of the city's rebound, they surely had a vested interest in contributing to the recovery efforts. Indeed, business interests sprang into action even before the fires were extinguished. Among the accounts in *Three Fearful Days* is that of cashier and manager Edward M. Lind, who describes the Herculean efforts to save thousands of gallons of Hotaling's whiskey at risk of going up (spectacularly) in flames. In editor Malcolm Barker's words, "divine intervention took human form in saving the whisky."[26] Springing into action while the fires raged, Lind rounded up 80 men with promised wages of $1 a day and organized a remarkable operation to rescue, move (roll), and guard dozens of barrels with their precious contents. Within three weeks of the temblor, A. P. Hotaling and Co. ran ads assuring their customers, "We have not been affected by the fire and are still doing business at the same old stand."[27] The ad did note that whiskey could be shipped only to addresses outside of the city: liquor sales remained banned in San Francisco for ten weeks following the earthquake. The ad also noted that only cash orders could be accepted, "as checks and exchanges are useless at present."[28] Still, the company was clearly and proudly open for business.

Elsewhere in the city, businesses sprang back to life wherever and however they could (figure 7.2). Colonel Edwin Emerson described a scene of energy and gaiety by October of 1906: businesses set up in new, low redwood shops, in deserted mansions, and in tents on what had been lawns and gardens. Emerson describes a city suddenly made new, "no shopworn goods or old stock left for the merchant to 'work off,'—they were disposed of without his consent."[29] Store awnings were new and clean, bright flags and banners flew from every storefront. Surveying the bright, bustling, chaotic scene, Emerson wrote:

FIGURE 7.2. Thriving free enterprise in the midst of the rubble.

All combines in one composite impression,—hurry and hopefulness, gaiety, silken petticoats and starched gowns, corduroys, tramping boats, sombreros, temporary wooden buildings, the flutter of many flags, rush of automobiles, clatter of lumber, banging of hammers and the rumble of drays,—the very sunlit air seems to breathe renaissance. (This is the spirit at the Golden Gate,—the old spirit that still lives— and this is why it is fun to live in the new San Francisco) [figure 7.3].[30]

The spirit of optimism expressed itself with exuberance in the January 1907 Los Angeles Feast of the Flowers parade. The city of San Francisco's entry that year: Fall and Triumphant Rise of San Francisco (figure 7.4).

San Francisco did not, of course, rise from the ashes overnight. Tent cities remained standing for many months, long enough for the rhythms of everyday life to be established. It would be more than three years before the city was ready to welcome visitors to return, for example, to the 1909 Portola Festival (figure 7.5). However, as summarized somewhat rapturously by Rufus Steele, rebuilding a city in three short years was no small feat. "Never did transformation crowd so closely upon the heels of devastation," Steele wrote. "After three short years one could stand upon the slopes of the Twin Peaks and gaze down across a majestic sweep of domes, towers, spires, and roofs, which stretches four miles to the Ferry without a visible break."[31]

The rebuilding of San Francisco brought with it modernization. Many of

FIGURE 7.3. Postcard showing the "new San Francisco."

the city's former cable-car lines—all but those that had to climb the steepest hills—were replaced with electric lines. For obvious reasons, the city tackled the job of establishing a safe and reliable water supply with hitherto unknown energy and determination. Chinatown got a face-lift, losing some of its authenticity and (perhaps) charm, but not too much. "There is less puzzle of blind alleys," Steele wrote, "more glitter of electricity, the same life under the ground."[32] Behind the glitzy new shop fronts the old flavors of Chinatown remained: "Roast pig still holds the sidewalk; there is still the fume of opium and firecrackers."[33]

The longer-term rebuilding effort grew, perhaps inevitably, out of shorter-term seeds of optimism: "the wine of the city's life has lost none of that 'flavor of the Arabian Nights' which was—is—body, sparkle, and bead,"[34] Steele observed, echoing sentiments expressed by so many others. But if optimism was the order of the day for San Francisco's citizenry and businesses in the immediate aftermath of the quake, what of the scientists and engineers of the day—John Branner and his colleagues, whose investigations were in danger of being thwarted by business interests? Without question the frustrations were real. Not many decades earlier, earth scientists had encountered similar pressures in their investigations of the October 21, 1868,

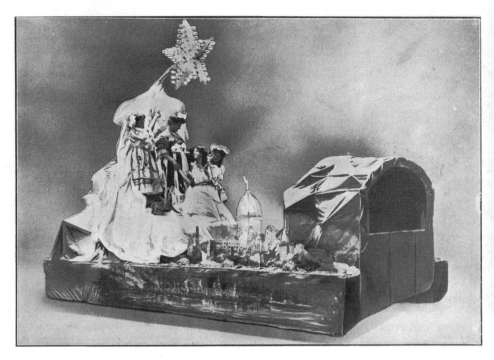

FIGURE 7.4. City of San Francisco entry in a Los Angeles parade: "Fall and Triumphant Rise of San Francisco."

Hayward fault earthquake. The nature of these pressures was showcased dramatically in a letter that George Davidson had written to the Seismological Society of America several decades after the earthquake. This letter was not published until 1982, when geophysicist Will Prescott published a note in the *Bulletin of the SSA*. In the letter, Davidson tells of the five subcommittees that had been formed following the 1868 temblor, three to investigate the performance of structures, one to summarize scientific results, and one devoted to legal matters. The news media eagerly awaited the report; news articles in January 1869 criticized the slow progress of the effort.

Among the findings of Davidson and his colleagues were some that are all too familiar to modern ears: most of the estimated $1.5 million in property damage had occurred on "made land"[35] and other sites where buildings had been built atop loose sediments and soils. In Davidson's words, "Report was carefully prepared but [committee chairman] Gordon declared it would ruin the commercial prospects of San Francisco to admit the large amount of damage and the cost thereof, and *declared he would* never publish it."[36] In

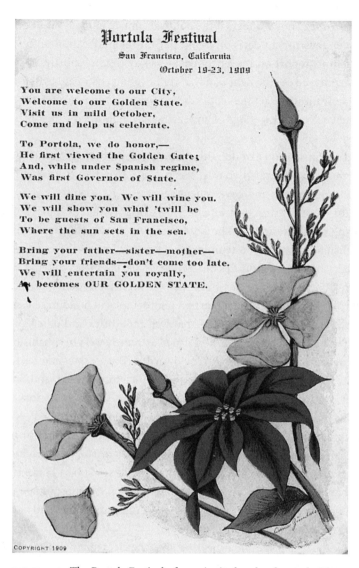

FIGURE 7.5. The Portola Festival of 1909 invited and welcomed visitors to the Golden State.

Davidson's estimation, "the earthquake of 1868 was more violent and destructive than that of 1906," the degree of damage due to the "suddenness and extent of the movement."[37] Regarding the missing report, Davidson added, with understated but obvious regret, "I know of no one who had taken a copy of the report"[38] prior to George Gordon's decision to suppress it.

In retrospect, one wonders if the political climate in 1906 was responsible for a later government decision to not pay for the publication of the extensive scientific report on the earthquake—as we discuss shortly, this earthquake report did see the light of day, its publication costs paid by the private Carnegie Institution. Yet the report did get written, and was one of the most comprehensive investigations of any earthquake published—before or even since 1906.

And as Hansen and Condon themselves note, in the three years following the earthquake, San Francisco was "overrun by researchers from engineering societies, universities and government agencies, who used the city as a vast laboratory to study structural engineering aspects of the earthquake and the following conflagration."[39] The best efforts of dreaded business interests notwithstanding, the 1906 temblor thus became one of the most exhaustively investigated earthquakes of all time. While engineers differed in their interpretations, a 1929 report published by the Clay Products Institute of California concluded, based in part on photographs of buildings taken after the earthquake but before the fire, that well-built brick and other masonry structures built on firm ground performed extremely well. In conclusion, this report agrees with the assessment of Edward M. Boggs, chief engineer of the Oakland Traction Consolidated and San Francisco, Oakland, and San Jose Railway, who stated, "In my opinion fully 95% of the property loss in San Francisco was due to the fire."[40] One might be inclined to take Boggs's words with a grain of salt: perhaps they were shaped by the same ulterior motive that inspired Horsburgh's disinformation campaign. The Clay Products Institute report, spearheaded by the Institute's president, Robert Linton, clearly could have had its own more general agenda. Yet a not insubstantial number of photographs taken before the fire do reveal masonry homes and buildings looking remarkably unscathed (figure 7.6).

While the conclusions may be debatable, the fact remains: the earthquake most certainly was investigated, by engineers and scientists alike. Not only that: for the field of seismology, the great San Francisco earthquake was a watershed event. By 1906 previous earthquakes had helped to launch both seismology and the modern era of thorough scientific investigation of and reporting on important earthquakes. In the decade or two that preceded 1906, the efforts of a few dozen scientific pioneers had also launched the field of *seismometry,* the design of sophisticated instruments capable of recording the waves produced by earthquakes. The 1906 earthquake did not launch the age of modern seismology, but arguably it was the first great earthquake *of*

FIGURE 7.6. Photograph taken after the earthquake but before the fire. Note the absence of obvious damage to even weaker architectural elements of buildings. (NISEE photograph)

the age of modern seismology and seismic recording. We now move on to discuss the earthquake in its scientific context—the scientific side of *elastic rebound.*

Elastic Rebound: A Science Comes of Age

American seismologists are sometimes inclined to point to 1906 as the birth date of seismology as a modern field of scientific inquiry. As a discipline, seismology is indeed very young, but it is not that young. As previous chapters have discussed, one can debate the beginning of the era of modern seismology, but many scientists point to Robert Mallet's pioneering treatment of earthquake waves as the starting point of investigations that present-day seismologists would recognize as akin to their own research. Also as previous chapters have discussed, notable earthquakes inspired scientific inquiries and even thorough reports decades before 5:12 on the morning of April 18, 1906.

Interestingly, while earthquakes had intrigued scientists and philosophers certainly since Aristotle's day, the earliest solid scientific advances in seismology concerned the waves that earthquakes generate rather than the nature of earthquakes (i.e., fault rupture) or their fundamental causes. Scientists' understanding of waves thus predated their understanding of earthquakes as a physical phenomenon. What is an earthquake? If we define the word to mean not the shaking generated within the earth but rather the physical process that occurs on a fault, we are left with a question for which scientists had no good answer even half a century after Mallet's pioneering contributions.

Only at the very end of the 20th century did scientists begin to understand that earthquakes occur on faults. As discussed in earlier chapters, prior to this time, surface breaks and faults had been recognized, of course, and in some cases described with laudable accuracy. But scientists viewed such breaks as being among the numerous effects—landslides, slumping, ground fissures—*caused by* earthquakes. The recognition of fault rupture as the essence of the earthquake process did not occur until the very end of the 19th century. The 1891 Mino-Owari earthquake in Japan is generally recognized as the event that cemented scientists' understanding of fault rupture as the primary cause of earthquakes.

By the time of the publication of "The Destructive Extent of the California Earthquake of 1906" in 1907, scientists tended to interpret faults as a consequence of vertical stresses caused, for example, by sinking of blocks of crust. (We now know that horizontal stresses were responsible for the 1906 earthquake.) Otherwise, words such as those of Charles Derleth, Jr., tend to ring fairly true to modern ears:

> To produce . . . equilibrium the crust must give at its weakest point. In this way a crack or slit, or as it is termed in geology, a fault, is produced. Within the confines of California one finds a region of structural weakness, and as has already been pointed out, the State is marked by a number of long fault lines running along the foothills of the high mountain ranges in the Sierra region and along the structural valleys of the coast ranges. Slippings and adjustments of the crust have occurred along these fault lines many times in the remote past, and the

present evidences of geological faults and rifts are the accumulations of many past earthquake breaks.[41]

Translated simply: California is earthquake country.

Thus, within just a few years on either side of 1900, several key pieces to the earthquake puzzle fell into place, regarding both California specifically and earthquake science more generally. First there was the development of seismometers to make faithful recordings of the waves generated by earthquakes. Second was the aforementioned understanding of the relationship between faults and earthquakes. A third critical leap would be inspired directly by the 1906 San Francisco earthquake, which left a spectacular surface rupture draped like a ribbon across the California landscape. This earthquake led Harry Reid to propose one of the most basic tenets in seismology: that earthquakes occur because elastic energy is stored over time along faults, and released in abrupt episodes of rapid fault motion (figure 7.7). This idea was not wholly new in 1906—sometimes one wonders if any idea in science is ever wholly new. In particular, pioneering geologist G. K. Gilbert had formulated the concept of an earthquake cycle in the 1880s. Gilbert's work presaged the concept of elastic rebound, but it fell to Reid to develop a mature scientific theory.

Elastic rebound occurs because while the earth's crust is to a good approximation rigid, it is able to bend without breaking, just as a flexed pencil will warp before finally giving way. Where forces drive blocks of crust in different directions, most notably at the planet's active plate boundaries, energy (or *strain*) builds up because, for the most part, large faults are not able to move freely. Whereas the flexure of a pencil is obvious, the flexure of the earth's crust is very subtle. Only in very recent times have scientists been able to observe this flexure directly with GPS technology. Elastic rebound might be a subtle phenomenon in some ways, but its consequences are anything but subtle: when faults give way and blocks of crust rebound, this snapping of the earth's fingers generates waves that lay waste to cities and literally ring the earth like a bell.

At the end of the 19th century and the beginning of the 20th century, one key puzzle piece was still a half-century away from discovery: the plate tectonics revolution, during which scientists at last understood the answer to a question that had intrigued the world's scientific minds for millennia: *Why* do earthquakes happen? Earlier theories involving underground winds, chemical mechanisms, electrical disturbances—all of these gave way to the elegant

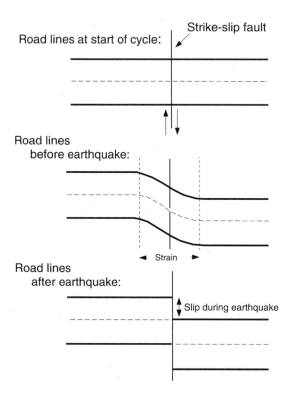

FIGURE 7.7. Illustration of the basic principle of elastic rebound theory. In the neighborhood of a fault, strain builds up while the fault is locked (middle), then releases during an earthquake.

simplicity of tectonic plates, their motion, and the energy that is stored at their boundaries.

Thus scientists' understanding of earthquakes was very much backward, from a modern pedagogical point of view. Scientists understood first the waves that earthquakes produce, and then the inherent process at work during an earthquake, and finally the forces that cause earthquakes to happen. This progression is interesting to consider. If one were explaining earthquakes as a physical phenomenon today, the inclination would be to proceed in the other direction: plate tectonics first, then the basic tenets of earthquakes, then the nature of wave propagation. Historically, however, if the earth were a dog, scientists had developed a good understanding of the tail's wagging well before they began to understand the nature of the tail itself, let alone the nature of the dog. Reid recognized the limitations of his theory: "The reasoning so far has been strictly along dynamic lines and the results may be accepted with some confidence; but in attempting to find the origin of the forces which produced the deformation we have been studying, we pass into the realm of speculation."[42]

Thus 1906 is neither the birth date of seismology as a modern field of inquiry nor the date that the original cause of earthquakes was understood. But it was a pivotal moment. The San Francisco earthquake played a tremendously important role in the development of understanding the tail, if not the dog. Earthquakes do not, of course, investigate themselves; rather, they provide key data that, if properly collected and analyzed, provide the grist for scientific advances.

The definitive report on the 1906 earthquake is generally known throughout the field as the Lawson Report, after Andrew Lawson, then chair of the Department of Geology at the University of California at Berkeley. The governor of California, George C. Pardee, had appointed a committee three days after the April 18 temblor to investigate its effects; Lawson served as its chairman. A preliminary report on the earthquake was submitted to the governor on May 31, 1906. However, while the government had given the committee its marching orders, it had, for the perhaps nefarious reasons mentioned earlier, not seen fit to provide financial support for either the committee's efforts or its publications. The privately supported Carnegie Institution in Washington, D.C., stepped in with financial support, and the committee's work continued.

The original report eventually grew into two volumes, the second a compendium of maps and reproductions of seismograms. The first volume was more than 400 pages, including several stand-alone papers by individuals whose authorship is indicated throughout the report. In addition to Lawson, several of seismology's founding fathers were among the committee's ranks: G. K. Gilbert, H. F. Reid, H. O. Wood, F. Omori, and others—names that are found on the pages of other chapters in this book.

The compilation of observations and interpretations in the Lawson Report—formally titled *California Earthquake of April 18, 1906*—is remarkable in both quality and quantity. The publication was perhaps overly parochial. In his book *The Founders of Seismology*, Charles Davison wrote: "No report on any previous earthquake has been issued on so liberal a scale."[43] However, Davison also observed a lamentable absence of reference to earlier earthquakes. "Indeed," he wrote, "if the Californian earthquake were the only shock known to mankind, the attention paid to it could hardly have been more exclusive."[44] Still, one cannot deny the important role of the 1906 earthquake, and the Herculean efforts of the committee, in advancing the field of earthquake science. The nature of earthquake ruptures, the standard for post-earthquake investigations and reports, the value of seismograms for investigating earthquake waves—all of these things were either established (or very nearly established) with the publication of the Lawson Report. And remarkable as this list is, it omits a fourth key item: the Lawson report documented the characteristics of California's most remarkable fault.

Scientists of the day did not recognize the San Andreas as a major plate boundary fault, of course; the word "plate," in a geologic context, would have been entirely alien, since the usage would not be coined for another half-century. Even Alfred Wegener's early theories of continental drift were decades away. But the 1906 earthquake left such a conspicuous scar across the land-scape that some of the founding fathers of modern geology and seismology were able to trace it for hundreds of miles—and to keep going for hundreds of miles more, mapping nearly the full extent of the San Andreas as we know it today. They didn't understand it fully but they did succeed in mapping it, and in so doing, they set the stage for generations of geologists and seismologists who were to follow. The remainder of this chapter will focus on this aspect of the 1906 earthquake: the role that it played in helping earth scientists to find faults in California.

The San Andreas Fault

Small parts of the San Andreas fault, in southern as well as northern California, had been recognized prior to 1906. In 1895 Lawson identified and wrote about a remarkably straight fault segment that ran through the San Andreas and Crystal Spring valleys south of the city of San Francisco. His initial writings reveal no hint that this feature might be part of something much larger. G. K. Gilbert, the man who had first used geologic observations to draw inferences about prehistoric earthquakes, was the first to describe the extent of the 1906 earthquake, and the amount of motion that had occurred, in his 1907 publication "The Earthquake as a Natural Phenomenon."[45] While Gilbert described the fault, he did not refer to it by name. The name appears to have been given by Lawson—whether it was given in honor of the San Andreas Valley or of himself must be left as an exercise for the reader.

Gossipy considerations aside, the initial pages of the Lawson Report lay out a truly remarkable account not only of the 1906 rupture but also of the San Andreas fault over nearly its entire extent. This account is all the more noteworthy when one remembers that the very idea of a fault rupture, and the relationship between faults and earthquakes, had only barely taken shape at that time. Lawson describes the extent of the San Andreas rift: in text and maps alike, the word "fault" is generally reserved for the 1906 rupture. Today earth scientists generally reserve the word "rift" for places where the earth's crust is being pulled apart, as is occurring today in eastern Africa. (Perhaps ironically, the usage "Dead Sea rift" has persisted even in modern times, in spite of the fact that this conspicuous linear feature is a fault, not a rift.) A "fault" is any surface on which earthquakes have occurred, and for any particular earthquake one speaks of a "fault rupture."

But Lawson and his brethren can scarcely be . . . well, *faulted* for their word usage. Whatever they called the rupture and the fault, the important thing is that they got it right. Teams of geologists traipsed over hill and over dale to map the fault break over its entire extent. The magnitude of this accomplishment is in only small part highlighted by the observation that the rupture was nearly 480 kilometers long. If one glances through the series of photographs in the Lawson Report, the accomplishment can be appreciated more fully. In 1906 California was not crisscrossed by a network of highways. In many parts of the state there were no roads. The first Model T was still two years away; the first Model T assembly line would not appear on the scene for another eight years.

It is fascinating to a modern Californian to consider some of the photographs in the Lawson Report and in mind's eye picture those locations as they look today. In 1906 most of California was a vast wilderness, punctuated by a relatively small number of cities with lots of open space—much of it rugged—in between. Even today it is not easy to trace the San Andreas fault from its northernmost land terminus near Point Arena to its southern termination near the Salton Sea. The fault winds its way through extremely rugged terrain in some regions. Along one segment through the Santa Cruz Mountains, poison oak is far more abundant in proximity to the fault than are roads of any size.

It is also interesting to the modern earth scientist to consider the organization of the Lawson Report. Following a brief introduction, the report turns to a description of the rift over its full mapped extent: not the 480-kilometer 1906 rupture, but the entire San Andreas rift, which Lawson and his cohorts traced for hundreds of kilometers south–southeast of the termination point of the 1906 rupture. Only after the report describes the full features of the rift does it move forward, in the next section, to discuss the details of the 1906 rupture.

Not surprisingly, mapping the rift and 1906 rupture was a committee effort. Dr. H. W. Fairbanks is credited by Lawson with having "kindly examined the ground"[46] from Point Arena to Fort Ross. Fairbanks traced the rupture and noted a difference in rock type across the fault in some locations: sandstones to the west, a formation known as the Franciscan to the east. The Franciscan formation has been well known to many generations of geology students in California. It is what geologists call a *mélange*: a geologic stew pot of different rocks and minerals. This stew represents essentially the detritus of plate tectonics: the remains and scrapings of old oceanic plates that were subducted in earlier times beneath the North American continent. The types of rocks within the Franciscan formation vary, but in some locations they are characterized by an appealing forest green hue and curiously slippery feel.

The contrast in rock type probably struck Fairbanks as a noteworthy curiosity. Later generations of scientists, however, would look to this type of observation to draw the inference that a tremendous amount of motion has occurred along the San Andreas fault—hundreds of miles of motion—over the geologic past.

South of Fort Ross the San Andreas makes a brief excursion out to sea before returning to dry land at Tomales Bay. G. K. Gilbert described the fault from this point south to Bolinas Bay, where it once again takes leave of the land. With acumen not surprising to anyone familiar with Gilbert's remarkable legacy of accomplishment, he describes the "rift" in some detail and discusses the relationship between the fault and the landscape through which it runs. Gilbert noted a series of ridges aligned parallel with the overall rift valley and observed that they disrupted drainage patterns of streams leading into the central Olema Creek. He also noted "local depressions, with ponds or small swamps," along linear valleys with the same overall alignment. He observed that these features have a character "so pronounced that forty-seven such ponds were seen between Papermill Creek and Bolinas Lagoon, a distance of 11 miles."[47]

Having far too penetrating an intellect merely to document such striking observations, Gilbert went on to explain them. "Their true explanation," he wrote, "is suggested by their relation to certain of the earthquake phenomena of April, 1906."[48] He observed that the rift comprised a number of fault strands, which experienced what Gilbert called "step-faulting," or relative vertical motion. He concluded that the ridges and sags within the rift zone were caused by uneven settling of small, narrow blocks within the larger zone of lateral (shear) motion.

Gilbert's explanation rings true to modern ears. Lateral faults are known to produce linear ridges where lateral motion brings a higher part of the fault into contact with a (formerly) lower part of the fault, or where bends or kinks of a fault produce a small component of compression that causes ridges to grow. Sag ponds are also generated at fault bends or kinks: whether such ponds are locally depressed or raised by lateral faulting depends on the direction of the fault bend relative to the direction of fault motion (figure 7.8). Gilbert's recognition of small-scale features and their relationship to faulting displayed extraordinary acumen, and helped set the stage for the modern subfield of geomorphology.

Photographs of the northern segment of the 1906 rupture reveal a fault winding its way through rugged, mostly uninhabited terrain. At Bolinas Bay there is a first hint of the earthquake's human toll: houses along the edge of the bay in the village of Bolinas, tossed about (in some cases into the bay) like Monopoly game pieces. Interestingly, the houses appear to have been sturdy, their shapes more or less intact even where the ground beneath them clearly gave way and slumped toward the bay.

The San Andreas fault crosses San Francisco Bay a few miles west of the Golden Gate Bridge. It reappears on land at Mussel Rock, about 13 kilome-

FIGURE 7.8. In tracing the San Andreas fault south of the 1906 rupture, geologists followed fault-generated features such as this sag pond along the fault near Frazier Park, north of Los Angeles. (Susan E. Hough)

ters south of the bridge. Lawson, along with Harry Wood, mapped the rupture from Mussel Rock southward along the San Francisco Peninsula, including the segment through the Crystal Springs and San Andreas valleys that Lawson had identified a decade or so earlier. One cannot help but wonder: Did he feel a sense of vindication, that he had identified a feature that proved to be so important, or a sense of chagrin, for having failed to appreciate its importance? In either case, Lawson's name is the one attached to the "Geomorphic and Geologic Map of the San Francisco Peninsula," published in 1907, that shows the trace of the fault—as well as a few other faults—from Mussel Rock to the Pajaro River.

The end point of Lawson's mapping efforts was judiciously chosen. To the south of the Pajaro River, along modern state highway 129, the fault, and the 1906 rupture, continue through some of the most rugged and inhospitable terrain that the San Andreas fault traverses. In keeping with decades of tradition governing the pecking order of academic science, Stanford University geologist J. C. Branner dispatched his beginning master's graduate student, G. A. Waring, to crawl over hill and dale, and through boundless tangles of poison oak, to map this part of the fault. Scientists now understand that this part of the fault is arguably the most complicated, and therefore the most scientifically interesting, segment of the 480-kilometer-long 1906 rupture. In 1906, however, it was probably considered mostly a nuisance, the documentation of which could safely be left to an individual with abundant time and, presumably, strong limbs. As a result, scientists today are left with the least expert, most sketchy observations in one of the places we care about most.

Waring's observations were, moreover, enigmatic; they included one photograph clearly showing that the far side of the fault had moved *left* relative to the side on which the photographer was standing. If one looks across a right-lateral fault (in either direction), the opposite side of the fault moves to the right, and the San Andreas is a quintessential example of a right-lateral fault. Could the negative have been reversed in the printing of Waring's photograph? Modern earth scientists had to wonder. When the 1989 earthquake struck in the same area, it generated similar left-lateral motion in the same location that Waring visited, thus providing the hapless graduate student a measure of posthumous vindication. Only in 1989 did scientists understand that, in 1989 as well as 1906, left-lateral motion had occurred on a secondary fault—not on the fault that was responsible for the main shocks. Waring's vindication was only partial: he had missed the San Andreas itself. In a paper published in 1991, geologists Carol Prentice and David Schwartz identified the San Andreas fault through the Santa Cruz Mountains. By 1998, Schwartz and his colleagues had found evidence for large prehistoric earthquakes along this segment of the fault.

The San Andreas fault and 1906 rupture emerge from the wilds of the Santa Cruz Mountains about where the Chittenden Bridge crosses the Pajaro River, along highway 129. Fairbanks mapped the fault from this point southward. Between Chittenden Bridge and San Juan Bautista (in 1906 simply San Juan) he observed a fence that had been broken and moved just over 1 meter. He then went on to San Juan Bautista, which he observed was standing "upon a bench of gravel," the northeast side of which "present[ed] a steep face."[49] At the site of the mission, the face was some 15 meters in height (figure 7.9). owes its existence to fault movement uplifting and tilting towards the southwest a portion of the floor of the valley, and that it thus originated in the same way as other similar features which we shall find to be characteristic of the Rift."[50]

From San Juan Bautista, Fairbanks's fault tour continued as he observed the rift as it follows the eastern slope of the Galiban (formerly Gavilan) Range. He described the rift as characterized by small valleys and gulches, and followed almost continuously by a wagon road. Perhaps the existence of the road beckoned him to continue his investigations. Fairbanks would not have been motivated by observations of fresh surface breaks, for San Juan Bautista marked the southern terminus of the 1906 rupture. In his discussion of the rift, Fairbanks does not note the disappearance of what he and his colleagues considered the *fault*. Thus did a report on one earthquake grow vastly beyond its obvious scope to become a seminal voyage of discovery of the San Andreas fault as a whole.

And so Fairbanks's tour continued through a part of California that remains thinly populated even today: the corridor along state highway 25, which passes through small towns such as San Benito and Bitterwater. Along this part of the rift, Fairbanks observed linear valleys, abrupt ridges, and markedly disrupted drainage patterns. He wrote that the rift could be traced (apparently easily) through the hills at the head of the Cholame Valley "by its characteristic features, as well as by bluffs which are undergoing rapid erosion."[51]

Here again, a geologist at the turn of the twentieth century did not know enough about the nature of faults in general, and of the San Andreas fault in particular, to begin to fully understand what he was seeing. (When speaking about the turn of the last century it is not necessary to write "he or she." To

FIGURE 7.9. Mission San Juan Bautista sits immediately atop a scarp along the San Andreas fault, at the southern terminus of the 1906 rupture. The mission was damaged by the earthquake but not catastrophically. (Susan E. Hough)

a nearly perfect approximation, the field geologists of the day were male.) The modern seismologically knowledgeable reader, however, instantly places the words of the Lawson Report into a well-established scientific context. South of San Juan Bautista the San Andreas fault is known to be creeping, which is to say that this part of the fault slips (more or less) steadily, never storing enough strain energy to generate large earthquakes. The fault zone is also especially straight and simple along this segment. The above two facts are probably not a coincidence. Scientists do not fully understand why a small minority of fault segments are able to creep, but the absence of geometrical complexities may well have something to do with it.

Fairbanks had little trouble following the straight course of the creeping segment. When he reached the small town of Parkfield, in the upper Cholame Valley, he made one of the most prescient observations found anywhere in the 434 pages of the Lawson Report. "The region around Parkfield . . . ," he wrote, "has been subjected to more frequent and violent disturbances than almost any other portion of the entire Rift."[52] Today, of course, we believe this to be (more or less) true: the observation of unusually regular moderate (magnitude ~6) earthquakes near Parkfield led to the U.S. Geological Survey's Parkfield Prediction Experiment. Monitoring efforts at Parkfield paid off in spades on September 18, 2004, when the latest Parkfield earthquake occurred.

In 1906 Fairbanks would have had no inkling of the prominence that Parkfield would attain in earth science circles, but drew his conclusions based on what he saw etched in the landscape around him. He described a prominent bluff nearly 60 meters high and, observing that the bluff was deeply eroded, concluded that it must have formed during "one of the oldest disturbances." In the low-lying region between this bluff and Cholame Creek, Fairbanks found evidence of "the effect of great disturbance over a considerable area." He wrote of parallel lines of faulting, "probably made at a later date than the bluff itself."[53]

Here again one cannot fully appreciate the degree of scientific acumen without pausing to consider the context. At the time of Fairbanks's wanderings and musings, geologists not only had only a nascent sense of the nature of faults, they also had only barely begun to understand how surface fault features could reveal evidence of past earthquakes. Fairbanks and his colleagues were piecing together an enormous and complicated jigsaw puzzle with only a glimmer of an idea of what a jigsaw puzzle was.

As Fairbanks's fault-finding tour continued south of Parkfield, his very first observation was "The people living along the Rift for 150 miles southeastward from the Cholame Valley tell wonderful stories of openings made in the earth by the earthquake of 1857. The first settler in Cholame Valley was erecting his cabin at that time, and it was shaken down."[54] The modern seismologically inclined reader again understands the context immediately. Parkfield marks the southern end of the creeping segment of the San Andreas fault and the northern terminus of the 1857 rupture.

The 1906 earthquake made it eminently clear that the northern segment of the San Andreas fault, from San Juan Bautista to Point Arena, can rupture of a piece. Although sketchy, available evidence suggests that smaller sections of the fault can rupture as well. Scientists understand, meanwhile, that earthquakes on the southern San Andreas fault will extend no farther north than the southern end of the creeping section: Parkfield. The only question is, how far southward can they extend from this point? In length and magnitude, the 1857 earthquake was nearly a twin of the 1906 event, extending some 350–400 kilometers.

Through the southern Cholame Valley, Fairbanks traced the rift into the Carissa Plain, known now as the Carrizo Plain, after the Spanish word for "reed grass." Again with almost eerie prescience, Fairbanks wrote of "[coming] upon the Rift at the northern end of the Carissa Plain, 4 miles northeast of Simmler P.O."[55] Today the tiny town of Simmler is at the base of the Temblor Range at state highway 58. "Here," Fairbanks continued, "the width of the broken country is much greater than usual, being nearly a mile. A number of lines of displacement can be distinguished; some nearly obliterated, others comparatively fresh. This is a region of light rainfall and gentled, grass-covered slopes, presenting just such conditions as would preserve for hundreds of years the effects of moderate displacements."[56]

So astute and careful were Fairbanks's observations that it sometimes seems impossible to improve on them: "The Rift zone continues to be traceable along the western base of the Temblor Range, finally passing out on to the gently rolling surface of the eastern edge of Carissa Plain. Broken, irregular slopes, cut-off ridges, blocked ravines, and hollows which are white with alkaline deposits from standing water mark the Rift. Carissa Plain has a length of about 30 miles."[57] Many decades later, geologist Bob Wallace would return to this section of the San Andreas fault to conduct his landmark investigations of past earthquakes along the fault, focusing on a series of features including a prominent offset streambed that Kerry Sieh would later name in his honor: Wallace Creek.

Most of the famous air photos of the San Andreas fault are taken along the Carrizo Plain (figure 7.10). In this location more than any other, a bird's-eye view conveys the sense of the fault as a scar across the landscape. One wonders if geologists of Fairbanks's day had any inkling of what the fault might look like from above. Did they ever daydream of joining the red-tailed hawks soaring high above the ground? As a geologic tourist, one certainly does so today—or, at least, from the ground level one's thoughts run to joining friends who have pilot's licenses and small planes at their disposal. Having seen photographs of the fault from above, geologists today are all too keenly aware that the scale of a major fault zone is well beyond the scale that is readily accessible, and digestible, by humans at ground level. In 1906, however, Fairbanks and his colleagues could not have imagined that they were following a boundary between two of the largest tectonic plates on earth, and so perhaps their eyes did not turn upward with longing. Or perhaps they

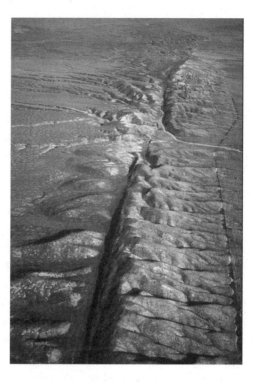

FIGURE 7.10. Viewed from the air, the now-famous Carrizo segment of the San Andreas fault was mapped in the aftermath of the 1906 earthquake by geologists who followed the trace of the fault from the ground, without the benefit of views such as this one. (Photograph by Robert Wallace; courtesy of U.S. Geological Survey)

did. In the proper tradition of scientific discourse, the contributors to the Lawson Report focused on their observations and, to a lesser extent, their interpretations. Philosophical musings are not the stuff that scientific reports are made of, then or now. One cannot help but wonder, though, what Fairbanks and his colleagues were thinking—and maybe discussing in the evening, over a beer, after long and arduous days in the field.

Or perhaps they were mostly sleeping. It would be hard to blame them.

Reaching the end of the Carrizo Plain, Fairbanks described a series of hills that "at first [seem] little more than a succession of ridges or hills cut off on the side next to the level plains. These detached ridges finally become connected in a regular line of hills with a steep but deeply dissected slope towards the southwest and long gentle slopes towards the northeast. This ridge is clearly a fault block, and now separates the southern arm of Carissa Plain from Elkhorn Plain."[58] As discussed earlier, geologists now know that ridges form along faults where bends introduce a component of compressional forces. The gradual appearance of ridges at the southern end of the Carrizo Plain marks the beginning of a bend in the San Andreas toward a more easterly strike.

Below Elkhorn Plain the fault cuts through terrain more rugged than the gentle contours of the Cholame Valley and Carrizo Plain to the north. Fairbanks observed that the rift was barely distinguishable in Bitter Creek and Santiago canyons, but found ample evidence of its handiwork in this region. "Huge masses of earth and rock are still moving," he wrote, "as shown by fresh cracks and leaning trees."[59] Fairbanks understood that rapid erosion and numerous landslides had obscured the trace of the rift through this region, yet he was able to follow its course through the area and beyond.

At Cuddy Valley, about 16 kilometers west of Interstate 5, the rift again revealed some of the features that by now must have been very familiar to Fairbanks: a linear valley, an abrupt fault scarp forming a steep bluff. From here Fairbanks traced the fault through Gorman Station. Photographs in the Lawson Report do not reveal much in the Gorman area besides fences and fault features. Today the region is among the most-visited sites along the San Andreas fault, although for the vast majority of visitors the distinction goes unnoticed. At Gorman the fault zone has carved a mountain pass along which Interstate 5 was built. In Fairbanks's day, the region around Gorman hosted not a superhighway but textbook examples of fault features. Fairbanks wrote of "a wonderfully regular ridge forming a marsh. In this vicinity the earthquake of 1857 is reported to have done much damage, shaking down an adobe house and breaking up the road."[60]

Fairbanks traced the rift as a series of valleys from Gorman Station to Palmdale. One can retrace his steps to a large extent today along Elizabeth Lake Road, which derives its name from Lake Elizabeth. Fairbanks observed, "Lake Elizabeth and Lower Lake are both due to the blocking of the drainage of two valleys extending along the Rift."[61]

Among earth scientists today Fairbanks is less well-known than a few of the other luminaries who contributed to the Lawson Report. Lawson himself, of course, is immortalized by his report, and maybe by the San Andreas. Harry Reid is well known as the developer of elastic rebound theory. But in following Fairbanks's words and footsteps, one cannot but be struck by his observational care and scientific acumen. Time and time again, he anticipated later developments that he could not have begun to understand, given the state of earthquake science during his lifetime. And many of his most astute inferences involved not the 1906 rupture but its larger geologic setting.

Following the rift beyond Lake Elizabeth, Fairbanks remarked on the nature of the rift from Leones Valley (now Leona Valley) to the point where it crossed the Southern Pacific Railway at Palmdale. Here, he wrote, "Movements have evidently been so often repeated and so intense along the Rift as to grind up the rocks and produce an impervious clayey stratum, bringing to the surface the water percolating downward thru the gravels of the waste slopes."[62] Today we know this clayey stratum as fault gouge, which is indeed the tailings of repeated motion and grinding along major faults.

In the vicinity of Palmdale, Fairbanks wrote of especially dramatic escarpments, as high as 12 to 15 meters in some places. Even by 1906, the natural lake within this depression had been enlarged for use as a reservoir. Today, Lake Palmdale is a major reservoir that can be viewed in full splendor from a scenic outlook off of Interstate 5 just south of Palmdale.

From Palmdale, Fairbanks's route along present-day Fort Tejon Road and Big Pines Road, followed the rift to an altitude of nearly 2,100 meters (near the present-day town of Wrightwood and its nearby ski resorts) and then down again along Lone Pine Canyon, which he observed to be "remarkable for its length and straightness." Today Interstate 15 offers a dramatic view up Lone Pine Canyon, just a few miles north of where it merges with I-215. Fairbanks wrote that "The uniformly straight course which the Rift exhibits in this portion of its length takes it diagonally across the mountains from the northern and desert side of the San Gabriel Range to the southern side of the San Bernardino Range."[63] Today scientists understand the role that the San Andreas has played in carving out what we now call Cajon Pass.

Fairbanks was able to follow the rift past San Bernardino to a mile southeast of Oak Glen, a small hamlet known today for its you-pick apple and cherry orchards. That the region supports fruit orchards is no coincidence. As Fairbanks observed, here there are "large springs which issue upon the line of the fault."[64] As he had inferred correctly, considering the clayey materials elsewhere along the fault, the rift often served as a groundwater barrier at which subsurface waters made their way to the surface—beckoning to flora and fauna alike.

To the east of Oak Glen the trail became more difficult to follow—not surprisingly, for even in recent times earth scientists have debated the continuity of the San Andreas fault through this region. As Fairbanks noted, the fault through this region is obscured considerably by landslides and erosion. He was, however, able to pick up the trail once again at Whitewater Canyon, which can now be reached by taking the Whitewater exit north from Interstate 10 just west of Palm Springs. East of Whitewater one reaches the Coachella Valley—known in 1906 as the Conchilla Desert—in which Fairbanks observed "the bedrock [to be] completely buried by recent accumulations."[65]

Modern fault maps of the Coachella Valley reveal a San Andreas fault split into two strands: the Mission Creek fault to the north and the Banning fault to the south (figure 7.11). Clear surface expression of this fault can be found in any number of locations, including the kinds of ridges that Fairbanks had observed so commonly elsewhere along his journey. Considering the length and Herculean nature of his journey, however, and the fact that it could easily have been blazingly hot by the time Fairbanks reached the Coachella Valley, one is inclined to feel forgiving that he did not follow the rift through this region as well.

One is especially inclined to feel forgiving when one reads what Fairbanks did not observe directly, but rather inferred. He wrote of evidence—albeit in his estimation weak evidence—that the fault continued into the Salton Basin: "the presence of mud volcanoes and several small pumiceous eruptions near the center of the basin." He finally concluded that "it may be reasonably assumed, then, from the best of our knowledge, that the southern end of the

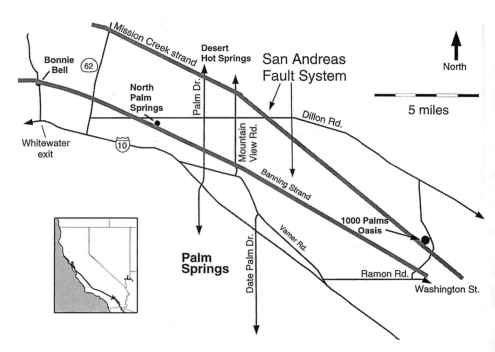

FIGURE 7.11. Map illustrating the complexity of the San Andreas fault system in the Coachella Valley, near Palm Springs. In the aftermath of the 1906 earthquake, geologists traced the San Andreas fault all the way down to near San Bernardino, but were unable to trace it through the Coachella Valley and speculated that it might instead follow a more southerly course, along what is now known as the San Jacinto fault.

great Rift is to be traced for an unknown length along the base of the mountains bordering the Salton Basin upon the northeast, in all probability gradually drawing out."[66] Here ended Fairbanks's own tour, and overview, of the rift.

At the end of this section of the Lawson Report, Lawson himself steps back in with "Review of Salient Features." He observes that the rift has been traced from Humboldt County to the north end of the Colorado Desert, a distance of over 960 kilometers. He notes that "thruout its extent the Rift presents a variable relation to the major geomorphic features of the region traversed by it."[67] The rift is observed to continue through mountains and alongside them, through valleys, and into deserts. Lawson notes the proximity of the rift to mountains along much of its extent, although the position of the former relative to the latter varies considerably. In this and much else, Lawson and his colleagues anticipated a great many of the later developments in earthquake science.

Earth scientists today understand a great deal about the San Andreas fault, an understanding that draws on a century of continued observations. In the 1940s Bob Wallace made the then somewhat scandalous assertion that the fault had moved as much as 112 kilometers over time. In 1953 Mason Hill and Tom Dibblee went even further off the deep end, presenting evidence for hundreds of kilometers of cumulative displacement. These studies preceded, and in some ways presaged, the plate tectonics revolution that followed not long thereafter.

Once plate tectonics theory established the San Andreas as the dominant player in the North American–Pacific plate boundary system, probably the next great advance in understanding came with the application of burgeoning ideas of paleoseismology. With encouragement from Wallace, graduate student Kerry Sieh first applied quantitative methods to investigate prehistoric earthquakes on the San Andreas fault. Within just a few decades of this work, geologists had made enormous strides piecing together the history and prehistory of the fault, in some places as far back in time as the birth of Christ.

Recent earthquake science, aided by technological developments such as GPS, computers, and C-14 dating, might have seemed to Lawson and his colleagues the stuff of which science fiction stories are made. But considering the state of the field of earthquake science when they found it, it is nothing less than astounding how much they did, and how much they did right, in their work to establish the character of the San Andreas fault. But perhaps even more remarkable, consciously or otherwise, Lawson understood that while it was remarkably long in its own right, the 1906 rupture had to be dis-

cussed in context, as part of a greater whole. At the end of his review he speculates about an even larger context. He writes: "The Colorado Desert and its
continuation in the Gulf of California are certainly diastrophic depression,
and may with much plausibility be regarded as a great Rift valley of even
greater magnitude than the now famous African prototype first recognized
by Suess."[68] He went on to note that "This great depression lies between the
Peninsula of Lower California and the Mexican Plateau. All three of these
features find their counterpart in southern Mexico."[69] Lawson wrote of a
pronounced "valley system" continuing to Salina Cruz, "well known to be the
subject of repeated seismic disturbances. On the same general line lies Chilpancingo, the seat of the recent disastrous Mexican earthquake. Following
these great structural lines southward, they take on a more and more latitu-
dinal trend; and beyond Salina Cruz the geological structure indicates that
this seismic belt crosses the state of Chiapas and Guatemala, to the Atlantic
side of Central America with an east–west trend, and falls into alinement
[sic] with Jamaica. It thus seems not improbable that the three great earthquakes of California, Chilpancingo, and Jamaica may be on the same seismic
line which is known in California as the San Andreas Rift."[70]

If christening the rift with the name San Andreas was an act of ego, in the
above conclusions ego overstepped its bounds. Scientists now reserve the
name San Andreas for a plate boundary fault that does end more or less where
Fairbanks supposed, at the end of the Salton Sea. The feature extends no farther because the nature of the plate boundary changes at both ends, in particular giving way to a mixed zone of lateral and extensional motion along
Baja California, and then subduction off the western coast of Mexico.

Lawson was absolutely correct, however, in his understanding—or at least
intuitive belief—that major rift systems connect. The line that he traced, extending nearly due eastward into the Caribbean, largely marks the boundary
of the North American and Caribbean plates (see chapter 9).

Here again one has to wonder. How far might Lawson's insight and
intellect—and, yes, ego—have taken him if he had had just a few more tools
at his disposal? In 1906, mapmaking was a matter of draftsmanship and real
craftsmanship; geologists such as Lawson would have been disinclined to
put together large-scale maps to summarize the observations that they documented with words. If he had had the ability to more easily generate maps,
displaying in graphical form the features and seismic belts that he inferred
to be connected, it seems likely that the nature of the connections would
have jumped off the page at him. Lawson and his colleagues had many of the

pieces to put together the plate tectonics puzzle; they lacked only the necessary tools to take the final step.

But what they could accomplish, they did accomplish. With the 1906 earthquake to set them on the trail, they described the San Andreas fault over nearly its full extent. Thus in a very real sense the 1906 earthquake led geologists to find fault in California—not just any fault but, although they couldn't know it at the time, the primary plate boundary fault of one of the most dramatic plate boundaries anywhere on earth. The fault is easily traced over some of its extent, especially once one knows where to look, but far more difficult to follow, especially on the ground, in other areas. Lawson and his colleagues managed to get it right, through thick and thin. They identified a number of regions where landslides had obscured the expression of the "Rift," but picked up the scent again on the other side. They observed the overall direction of the fault to be quite straight until Tejon Pass, at which point they documented the beginning of the one prominent bend along the fault. They were unable to identify the fault through the Coachella Valley but speculated that it must continue. They even mapped some of the San Jacinto fault, now known to be the important second actor in the southernmost San Andreas system. With this seminal field effort to build upon, geologist Bailey Willis published a fault map of California in 1923, one that bears a striking resemblance to modern fault maps of the state.

At the end of the journey, what is the story of the great San Francisco earthquake of 1906? A cover-up, clearly; a disaster, certainly, and one that could not be denied. The world saw a great city devastated, reduced to ash and rubble, and responded with compassion and aid. As Californians, as Americans, but most importantly as human beings, survivors of the temblors' greatest wrath responded as human beings have always responded to the most demanding trials: with resilience and deep wellsprings of optimism. Scientists of the day responded in kind, and in so doing advanced a nascent field of inquiry by leaps and bounds.

"Elastic rebound": These two words at once represent one of the most important tenets of earthquake science and the most fitting metaphor for the response of both man and mankind to the planet's most devastating disasters. A few short months after the fateful day of April 18, 1906, scientists were on the path of discovery and renaissance was in the sunlit air of the new San Francisco. A city reborn, a science come of age.

8

The 1923 Kanto Earthquake: Surviving Doomsday

> The critic will say that the old city was not worth restoration;
> that it was nothing but a feudal municipality, a vast
> congregation of villages patched with modern improvements;
> that it needed reconstruction, not restoration.
> —K. Sawada and Charles A. Beard, "Reconstruction
> in Tokyo" (March 1925): 268

Citizens of Yokohama and Tokyo were just sitting down to their Saturday noonday meal on the morning of September 1, 1923, when the great Kanto earthquake struck. The time, 11:58:44, was precisely documented by seismometers, which were by this time commonplace. In 1923, Tokyo was already a bustling urban center and port city, home to over 2 million people. Yokohama was an important port and industrial center as well, with a population of more than 400,000.

As had been the case in Charleston, observers gave differing descriptions of the initial shaking; some witnesses described the same gradual onset that residents of Somerville, South Carolina, had experienced. In Yokohama, however, Otis Manchester Poole wrote that, in contrast to other temblors that allowed time for contemplative speculation ("How bad is this one going to be?"),

This time . . . there was never more than a few moment's doubt; after the first seven seconds of subterranean thunder and creaking spasms, we shot right over the border line. The ground could scarcely be said to shake; it heaved, tossed and leapt under one. The walls bulged as if made of cardboard and the din became awful. . . . For perhaps half a minute the fabric of our surroundings held; then came disintegration. Slabs of plaster left the ceilings and fell about our ears, filling the air with a blinding, smothering fog of dust. Walls bulged, spread and sagged, pictures danced on their wires, flew out and crashed to splinters. Desks slid about, cabinets, safes and furniture toppled, spun a moment and fell on their sides. It felt as if the floor were rising and falling beneath one's feet in billows knee high.[1]

Poole could not gauge how much time elapsed during the tumult but cited an official record of four minutes.

Although the earthquake damaged all of the seismographs operated by the seismological station at Tokyo University, Professor Akitsune Imamura and his staff were at work within minutes of the earthquake, analyzing the seismograms. Within 30 minutes they were able to provide preliminary information about the temblor to the press. The seismograms revealed evidence of three separate shocks, the largest of which struck beneath Sagami Bay. Land surveys revealed myriad fractures and zones of coastal uplift, some as much as 2 meters high. All of these were mere secondary fractures generated by the primary fault rupture far beneath the floor of the bay.

The Kanto earthquake struck several decades before the plate tectonics revolution provided scientists with a clear conceptual view of Japan as an active plate boundary zone. By now scientists have learned a great deal about the subduction zone atop which Japan sits. Indeed, we know now that Japan owes its very existence to the subduction zone. The country is an amalgamation of island arcs, volcanic mountain chains that can develop offshore in regions where subduction begins beneath the ocean (figure 8.1). Thus Japan is exposed to far higher earthquake hazard than its neighbors to the west, which are much farther from the active plate boundary zones (see sidebar 8.1). Scientists at institutions such as the Japan Marine Science and Technology Center have used data from conventional seismometers as well as sophisticated ocean-bottom seismometers and other instruments to assemble detailed images of the three-dimensional geometry of the Japanese subduction zone, which represents one of the most active and dangerous subduction zones in

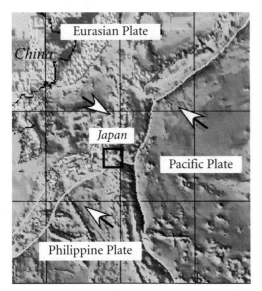

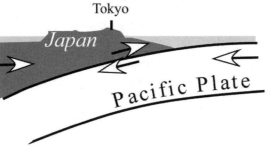

FIGURE 8.1. Geometry of tectonic plates in the vicinity of Japan. The Pacific plate sinks, or subducts, beneath the Eurasian plate, creating the island arc of Japan. The country owes its very existence to the active plate tectonic forces that now imperil its citizens.

the world. Along this zone, the rate at which oceanic crust sinks is among the highest in the world. The zone comprises several segments thought to be distinct in their geometry and production of major earthquakes, most notably (for the most populated parts of Japan) the Sagami trough and the Nankai trough. Great earthquakes have ruptured all or parts of both troughs in recent decades, including large Nankai trough earthquakes in 1944 and 1946. The 1923 earthquake is now known to have struck the neighboring Sagami trough.

In addition to the data from modern earthquakes and seismometers, scientists have drawn on Japan's appreciable written historic record to learn about earthquakes as far back as A.D. 684. (Scientists at the Geological Survey of Japan have, in fact, taken this analysis one step further, using historic records from Tokyo and Tohuku to learn about historic earthquakes in northeastern Hokkaido, for which little written history is available prior to 1800.)

SIDEBAR 8.1

The most straightforward plate tectonics cartoon illustrates why Japan is subject to such high earthquake risk. To the west, neighboring China is considerably more complex. Along its southern boundary China experiences earthquakes related to the collision between India and the Eurasian plate. But large earthquakes strike elsewhere in China as well, sometimes with devastating consequences. The 1976 Tangshan earthquake struck in northern China, just a stone's throw from Beijing. As is the case with the large earthquakes that pop up in "unusual" areas on other continents, scientists still struggle to understand why these temblors occur and to estimate the hazard from future "unexpected" events.

Along the Sagami and Nankai troughs the historic record is both complete and clear: great earthquakes, such as the 1923 temblor, will recur on average about once every 100 years.

Concepts such as recurrence rates were very much in their infancy in 1923. Although the association between earthquakes and faulting had been established by the 1891 Mino-Owari earthquake in Japan, the fault that ruptured in 1923 was underwater, inaccessible to direct observation. It nonetheless represented an important earthquake for seismology—for Japanese seismologists in particular. One of the founding fathers of seismology, Fusakichi Omori, was in Australia when the earthquake struck. He returned to Japan on November 8, but died shortly thereafter; the task of directing the scientific response fell to his colleague, Professor Imamura. The investigation of the earthquake included surveys of damage and other effects (such as land surveys of coastal disruptions) as well as studies of waves recorded on seismographs over distances up to 1,600 kilometers.

Perhaps most intriguingly, S. Haeno of the Japan Seismological Institute examined data from a pair of horizontal pendulums that recorded small changes, or tilts, in the ground. Haeno concluded that the ground had been tilting, slowly but perceptibly, in the week prior to the earthquake, achieving a maximum tilt of about 1/3 minute immediately prior to the earthquake. (As a measure of angle, one minute represents 1/60th of a degree.) Even today, measurements of such precursory changes prior to large earthquakes are rare—nearly nonexistent—and scientists debate whether they even exist.

Scientists of the 1920s cannot be blamed for not making sense of Haeno's intriguing results; scientists today cannot fully make sense of them either. They do, however, provide a tantalizing suggestion that the earth initiates a launch sequence prior to at least some earthquakes. And this, in turn, suggests that it might someday be possible to detect signs of the launch earthquake, and know that the temblor is on its way.

When the next large earthquake strikes central Japan, it will be very well recorded on monitoring instruments that now blanket the islands. This instrumentation includes strain and tilt meters as well as GPS receivers—all of which should provide data to confirm or refute Haeno's results, and bring scientists one step closer to figuring out if earthquakes will ever be predictable. Future large earthquakes will also be recorded on hundreds of modern seismometers and accelerometers, instruments specially designed to record very strong shaking. In their commitment to earthquake monitoring, both Japan and Taiwan—two countries for which earthquakes are a national concern—have far outpaced efforts in any part of the United States, where earthquakes are often perceived as a regional problem. (In stark contrast to the United States, Japan spends more on earthquake research than on national defense.)

The seismograms available to Professor Imamura and his colleagues in 1923 were unlike the recordings made by modern seismometers and accelerometers, which can faithfully record every shake, heave, and toss of the ground—and allow scientists to make direct, quantitative estimates of the duration of strong shaking. As always, reported estimates of shaking duration varied considerably: four minutes, ten minutes, as much as two and a half *hours* of constant motion. In the closing years of the 20th century, seismologists David Wald and Paul Somerville used modern methods to analyze recordings of the Kanto earthquake made on seismometers around the globe to estimate the spatial extent of the earthquake and the amount of slip on the fault. This analysis, which focused on seismograms recorded on instruments worldwide, led to a precise estimate of the magnitude of the earthquake: 7.9. According to their calculations, the rupture extends some 130 kilometers, and the actual fault motion lasted about 45 seconds. The duration of strong shaking would have been considerably longer, of course, because waves left the fault and reverberated around in the crust. The estimate of four minutes is not implausible. And while the main shock itself did not last for hours, more than 200 aftershocks were documented between noon and midnight on September 1, an astonishing rate of one every 3½ minutes. Considering that the larger aftershocks would have generated their own reverberations, it is not surprising that some witnesses described hours of constant motion.

If the Japan subduction zone were less complex, its largest earthquakes would likely be fewer in number but even larger in size. M8 temblors appear to be the norm along the segmented zone—not in the same league as the M9+ temblors that have struck Alaska, Chile, and Sumatra, but bad enough. The lesson of 1923 was clear: an M8 temblor beneath the heart of Japan is more than sufficient to wreak havoc on the country's structures and people.

For the most part the continued aftershock tremblings were unnerving, but not of significant consequence to life and limb or to structures. The initial few minutes of the main shock were responsible for the havoc that ensued. Over 140,000 lives were lost that day, 59,000 of them in Tokyo. Total property damage in Japan was estimated at more than $2 billion (1923 dollars). Adjusting those dollars for inflation, and considering photographs taken in the aftermath of the great earthquake, it is clear that damage caused by the Kanto earthquake far outstripped that caused by more recent temblors such as Northridge and Kobe.

In Tokyo, the Building Inspection Department recorded the extent of the damage in that city. In what must have been an exhaustive effort, they assessed the level of damage to structures and tabulated it according to building type. Its survey reveals some differences in the performance of different building types. Of reinforced concrete and steel structures, about 40 percent were heavily damaged, with only 1–2 percent of each type collapsing wholly or in part. Brick and masonry buildings fared worse, with 60–70 percent sustaining heavy damage. About 7 percent of brick buildings collapsed wholly or in part; 4–5 percent (one out of 20–25) of other masonry (dressed-stone) structures met with the same fate. Only 10–20 percent of buildings escaped unscathed.

Photographs of Yokohama evoke thoughts of a war zone, with an almost eerie resemblance to photos taken just 17 years earlier in San Francisco. In fact, the similarity is no coincidence: enormous, unchecked firestorms devastated both Tokyo and Yokohama in the aftermath of the main shock. Many of the fires were ignited by cooking stoves in use to prepare the midday meal; dozens of fires began on the immediate heels of the earthquake. The day was unusually hot and windy, and the winds grew stronger as the day wore on. By early afternoon, wind speeds in Tokyo were 24–28 miles per hour; by 5 o'clock they reached 30 mph. At 6 o'clock the wind direction changed abruptly and the wind speed intensified even further. By 11 o'clock that night, wind speeds reached nearly 50 mph.

Considering the effects of the fire and earthquake separately, later investigators came to the rather startling conclusion that the earthquake caused damage to the tune of about 10 percent of the total building valuation in the hardest-hit areas. According to a published study by Robert Anderson, the Tokyo Metropolitan Police tallied some 375,000 buildings destroyed by the fire and less than 10 percent of this number, 34,000, destroyed by the earthquake itself. Of the latter structures, over 32,000 were wood frame buildings. Wood frame buildings fare well in earthquakes as a rule, but not when they have especially heavy roofs—as did many traditional Japanese houses of the day.

The fires' toll on human lives was also steep. In the Hongo Ward in Tokyo, over 30,000 refugees had congregated in a large open area adjacent to the Military Clothing Depot. By late afternoon a large school across the river from this building was engulfed in flames. When the wind direction shifted abruptly at 6 o'clock, the fire jumped across the river. The firestorm engulfed and literally incinerated the refugees who had gathered in what they thought was a safe haven. According to reports—and documented by ghastly, ghostly photographs—not a single soul survived.

Along with many others, Otis Poole and his family sought safety on the water: "How wonderful it was to scramble on board the *Daimyo* and stow everybody away comfortably; and a stiff peg of whisky all round never tasted better." Even from a safe vantage point, Poole describes a sleepless night of horror:

> In the enveloping summer night, the relentless roar of flames sounded like heavy surf, with frequent crashes of thunder. We seemed to be in the centre of a huge stage, illuminated by pulsing, crimson footlights. . . . we could see a thin rim of fire all around Tokyo Bay, meaning that fishing villages and small towns were all sharing the same fate; the glare above Yokosuka, where the jaws of the bay come close together, showed that the Naval arsenal was also going up. Northwards over the water there rose on the horizon a billowy, pink cloud like cumuli at sunset, so distant as to seem unchanging and motionless, yet each time one looked it had taken a different shape. This was Tokyo burning, and by the cloud's titanic proportions we knew the whole city must be in flames, as indeed most of it was.[2]

To some extent the Kanto earthquake represents an engineering success story: modern, well-built structures stood their ground admirably in the face of fierce ground shaking. At the time of the earthquake, the Mitsubishi Company had built and still owned 135 buildings of various construction types; none of these well-built structures were seriously damaged by the earthquake. The earthquake illustrated the need to make buildings not only earthquake-proof but also more fireproof, using fire-resistant building materials and powerful sprinkler systems in large structures.

In the aftermath of any damaging earthquake it can be hard to separate fact from fiction. Following the Kanto earthquake, corporate self-interest fueled some of the latter. It was stated in a 1929 report published by the Clay Products Institute of California, "As was probably natural in the keenly competitive building industry of today, immediately following the Japanese disaster there was a rush by the proponents of each class of building construction to prove that theirs was the only material with a satisfactory record. Out of the mass of such claims it is difficult to arrive at the true facts."[3]

These swirling claims notwithstanding, without question one building in particular generated more myth and legend per square foot than any other: the Imperial Hotel, designed by Frank Lloyd Wright and, by happenstance, in the midst of its grand opening on the day the earthquake struck (figure 8.2). Baron Okura, a key financial promoter of the project, sent a radio telegram reported to be the first word of the disaster to reach the United States: "HOTEL STANDS UNDAMAGED AS MONUMENT OF YOUR GENIUS. HUNDREDS OF HOMELESS PROVIDED BY PERFECTLY MAINTAINED SERVICE. CONGRATULATIONS. OKURA."[4] (See sidebar 8.2.)

By the 1920s the peril of earthquakes in Japan was well understood, so Wright had followed established practice in Japan, designing his grand hotel with earthquake forces in mind. He concluded that overly rigid buildings were to be avoided in earthquake country, arguing, "flexibility and resiliency must be the answer. . . . Why fight the quake? Why not sympathize with it and out with it?"[5] Wright's idea of "flexibility" was not the same as the general usage of the word in earthquake engineering circles. Whereas flexibility usually describes the ability of a building to flex, or sway, and thereby absorb shaking energy without damage, Wright's idea of flexibility was largely a matter of modular rigidity. Recognizing the vulnerability of large and rigid structural elements, Wright designed, in his own words, a "jointed monolith": "Where the parts were necessarily more than sixty feet long," he instructed, "joint these parts, clear through floors, walls, footings, and all, and manage the joints in the design."[6] As pointed out by Robert King Reitherman in a paper published nearly 60 years after the earthquake, Wright's insight is recogniz-

FIGURE 8.2. Postcard of the Imperial Hotel in Tokyo, designed by Frank Lloyd Wright.

able to modern engineers, who today frequently employ the seismic separation joint in the design of freeway overpasses and other large structures.

Wright's design proved its quakeworthiness in an especially timely fashion: not many structures are tested by a great earthquake on the very day they open for business. Okura's telegram—for a while the only communiqué from Japan to the United States—led to the immediate misconception that all of Tokyo lay in ruins except for the Imperial Hotel. In fact the structure did sustain some damage: the dining room floor warped and had to be releveled, and nonstructural items such as fans and lights were damaged.

Damage to the hotel's dining room floor was caused by another design element that was ahead of its time, but in this case also fundamentally misguided in its implementation. Sizing up the muddy site on which the hotel was to be built, Wright concluded, "That mud seemed a merciful provision—a good cushion to relieve the terrible shocks. Why not float the building upon it?—a battleship floats on salt water."[7] Thus did Wright arrive at his novel design for the hotel foundation: tapering concrete piles just nine inches wide

SIDEBAR 8.2

Although the earthquake severed most of the communications channels between Japan and the rest of the world, the earth itself telegraphed news of the earthquake well in advance of any human-generated communiqué. By 1923 seismometers were a common fixture in seismological laboratories. Just a dozen minutes after the Kanto earthquake struck, the P waves from the powerful temblor reached the Seismological Laboratory at Georgetown University; the S waves, about 11 minutes later. Observing these waves, laboratory director Rev. Francis Tondorf estimated a distance of 10,080 kilometers to what he knew must have been a powerful earthquake. This estimate was furnished to the Associated Press and sent out over its wires three hours before anyone in Washington, D.C., received "word" of the disaster.

and eight feet long, set every two feet along the length of the wall. Decades later, engineers would revisit the notion of "base isolation": the construction of shock absorbers within the foundation of a building to cushion the impact of shaking. Wright's design, however, was misguided. The primary effect of a layer of mud is to *amplify* earthquake shaking. A layer of mud also lacks the fundamental stability that defines the essence of a good foundation. When the hotel was demolished in 1968, it fell victim largely to economic forces in a city where real estate had become too precious for sprawling, low-rise hotels. But foundation problems also contributed to the downfall of the structure, which continued to settle, sometimes unevenly. Between 1955 and 1965 it sank 12 centimeters; a back part of the hotel subsided by a whopping 1 meter throughout the lifetime of the structure. Although architecture aficionados worldwide protested the planned demolition, the owners had a leg to stand on when they deemed the building "impossible to repair" by virtue of its ongoing settling.

In the aftermath of the 1923 temblor, however, architects and engineers probably found a measure of satisfaction in the performance of structures built with modern design and sensibilities. Unlike early 18th-century Europeans, the early 20th-century Japanese had no illusion of the universe as orderly and predictable: they knew only too well how unpredictable and devastating earthquakes could be. But perhaps by 1923 the Japanese were comforted by a different paradigm: that by designing and building structures appropriately, earthquake risk could be successfully mitigated. This paradigm is no illusion; it forms the underpinnings of modern earthquake risk mitigation efforts. Even today, however, earthquake engineering remains an imperfect science; in 1923 it was barely in its infancy. The collective toll of the earthquake and subsequent fire was catastrophic. The earthquake sent literal and figurative shock waves through local infrastructure, damaging communications lines, railroad tracks, dams, sewers, and more. The web of a modern, industrialized society was left in tatters.

The earthquake tore at Japan's social fabric as well. Rumors of looting by Korean immigrants—at that time largely a population of migrant workers—stoked anti-Korean sentiments and led to outbreaks of sometimes lethal violence aimed at the Korean community. Japan's economy had fared fairly well in the immediate aftermath of World War I, during which Japan had remained neutral. But by the time of the Kanto earthquake, Japanese businesses had begun to be hurt by a worldwide postwar economic slump. The earthquake added enormously to Japan's economic woes. The Bureau of Social Affairs estimated that nearly half of all workers lost their jobs in the immediate aftermath of the earthquake; by 1924 the economy as a whole was in recession.

By some accounts the Kanto earthquake sent the Japanese economy into a tailspin from which recovery was measured in years. The Great Depression sent the global economy into a tailspin just half a dozen years later; it seems only logical that the former downturn contributed to the latter.

In recent years scientists have in general recognized a greater degree of *interconnectedness* of complicated systems—and systems don't get much more complicated than human societies. Mathematicians speak of sensitivity to initial conditions within so-called chaotic systems. Mathematicians in blockbuster Hollywood movies speak of the "butterfly effect." As first described by meteorologist Edward Lorenz, a butterfly flaps its wings in Brazil and sets off a chain of events that ends with a tornado in Texas. The present is the sum total of the innumerable small disturbances of the past. Most small disturbances remain small, but a few cascade in ways that are impossible to predict: the chance meeting that leads to a marriage that produces an offspring who turns out to be the next Albert Einstein. That sort of thing.

If small-scale disturbances can cascade into big impacts, clearly large-scale disturbances stand a better chance of affecting everything that follows. The hallmark of a chaotic system remains the same, however, whether a given disturbance is big or small: unpredictability. Given an enormously complex system, it is impossible to predict how any disturbance of any size will shape the future. It might be fun to play the "what if" games, as in: What if John F. Kennedy had not been shot? But if it's impossible to predict the future, it is equally impossible to predict the past with a different set of starting conditions.

Would mid-20th-century Japan have been different if the Kanto earthquake had never happened? Such questions fall into the realm of the unknowable. A scientist can, however, retreat to the reliability of cold, hard facts to address a presumably related question: What was the long-term impact of the Kanto earthquake on the Japanese economy? One cannot hope to answer this question fully, economic systems being every bit as chaotic as human societies. But in this case, per capita GNP figures presumably shed some light on the extent of the long-term impact. Such data are available for 20th-century Japan.

Economic and scientific data oftentimes have much in common—in particular, the small-scale bumps and jitters riding atop the longer-term trends. These bumps and jitters can allow a degree of artistic license in the interpretation of what scientists call *noisy data*. Artistic license can, in turn, introduce a degree of subjectivity into one's conclusions. Considering the GNP curve, one might be tempted to draw a straight line through the values between 1890 and 1920 and a second, much flatter straight line between 1920 and 1930 (figure 8.3). Indeed, the aforementioned early 1920s dip is apparent in the GNP figures, as is the dip immediately after the Kanto quake.

On the other hand, one could allow a single straight line to blast its way from 1890 all the way to the early 1940s. The real GNP data veer away from this line in both directions, of course, but the single trend is not such a bad description of the Japanese economy over this 50-year period. During this time, per capita GNP increased by 300 percent, or about 2–3 percent per year.

The real blip in Japan's GNP curve, of course, corresponds to the immediate aftermath of World War II. The second most conspicuous feature of the curve is its steep rise following this steep decline. Between about 1947 and 1970, the per capita GNP grew at an impressive clip, rising from a hair over $1,000 (in 1990 U.S. dollars) to almost $10,000.

In the physical sciences parlance, World War II was what we call a *signal* in the data. The abrupt downturn, for which an explanation is readily at hand, represents a significant departure from the trends over the previous half-century. The key word here is "significant." When interpreting trends,

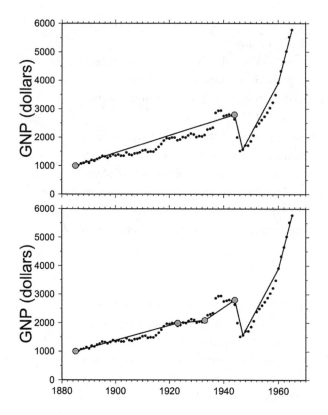

FIGURE 8.3. Both panels depict GNP in Japan during the years 1884–1964. Lines in the bottom panel indicate slowing of GNP growth in the years following the 1923 Kanto earthquake. Lines in the top panel indicate more or less constant long-term growth between 1884 and World War II, which caused a precipitous drop in GNP.

the scientist looks for signals that stand out among the normal bumps and jitters that typify noisy data. Returning to the prewar years, then, and considering the departures from the single straight line drawn from 1890 to 1944, one finds that none of the individual bumps and jitters stands out distinctly from all of the others. Both the Kanto earthquake and the Great Depression caused short-lived dips in the GNP figures, yet both dips were indeed only blips: no bigger than the usual sorts of fluctuations seen in the curve.

It may strain credulity that an earthquake as devastating as Kanto might generate an economic blip no larger than a run-of-the-mill fluctuation in the business cycle. Yet consider a few more basic numbers. With a population of about 56 million people in 1923, Japan's total GNP was about $110 billion. A loss of $2 billion amounted to less than 2 percent of the total GNP— a serious blow, to be sure, but, it seems, scarcely a fatal one.

The death toll, while staggering in human terms, represented about ¼ of 1 percent of the population. In Tokyo, some 59,000 lives were lost out of a population of over 2 million people. While staggering in human terms, in cold statistical terms even this loss was relatively small, less than 3 percent of the population.

Would the GNP curve have been different from 1923 onward had the Kanto earthquake never happened? Without question yes, but we don't know how. The story had played out previously in cities such as Lisbon, Naples, and Charleston; in 1923 it played out in Japan. Against formidable odds the cities of Tokyo and Yokohama were rebuilt in the years following the quake—and arguably, not merely a rebuilding but rather a rebirth. Prior to the earthquake both Tokyo and Yokohama had been something of a maze of narrow streets and small buildings. After the earthquake wiped the slates clean, Japanese capital flowed in, fueling a reconstruction along more modern lines: wider streets, parks, modern skyscrapers. Before the earthquake Tokyo was like a large, close-knit village in which most houses were built of wood, with light frames but often heavy roofs. The earthquake essentially wiped the landscape clean of structures.

In 1925, K. Sawada and Charles A. Beard published an article titled "Reconstruction in Tokyo." If the latter name sounds familiar, it is because, having founded the School of Politics at Columbia University in 1907, Beard went on to write numerous influential publications on local government. He was invited to Japan following the earthquake to contribute to the reorganization of the local government. Sawada and Beard, too, commented on the state of the city before the earthquake: "the critic will say that the old city was not worth restoration; that it was nothing but a feudal municipality, a vast congregation of villages patched with modern improvements; that it needed reconstruction, not restoration. Was it not a fatal error, an anachronism to rebuild the samurai city in the year of grace 1924?"[8] As Sawada and Beard went on to discuss, the "forces of 'business as usual' were pitted against the forces of progress," and to some extent progress lost out—for example, in the hasty construction of some temporary buildings (figure 8.4). But "beneath the surface, significant events were happening." Among the notable changes: city district plans were redrafted so that landowners retained at least 90 percent of their former lot size, but narrow, irregular, and sometimes blind alleyways gave way to regular and wider streets (figure 8.5). "The world will take note and long remember" Tokyo's planning efforts, Sawada and Beard predicted after reviewing efforts that were not yet complete but well under way by 1925.

Otis Manchester Poole described old Yokohama with obvious and wistful affection: "The Settlement presented an amiably haphazard array of offices, banks, fenced-in dwellings, stone godowns, tea-firing premises exuding a

FIGURE 8.4. Temporary buildings erected in the aftermath of the earthquake. (Page 268 from K. Sawada and Charles A. Beard, "Reconstruction in Tokyo," [March 1925])

dreamy fragrance, churches, restaurants and well-equipped livery stables, of one which ran stage-coaches to Tokyo until the first railway was built."[9] His book, tellingly titled *The Death of Old Yokohama*, concludes with a poignant reflection: "The Yokohama I still see in my mind's eye is the old one created by the pioneers, with its open bay and virgin hills I roamed in as a boy. Thoughts of those scenes and of the community I knew so well bring on a nostalgia that is like the scent of incense in a temple grove. In this I know that I am not alone."[10]

Poole would have likely been inclined to point out that "modernization" does not necessarily equate to progress or improvement, but much of Yokohama, including its harbor district, certainly was modernized after the quake (figure 8.6). Local authorities clearly recognized the opportunity: "Before permanent rebuilding on the old sites could render impossible any basic change," Poole wrote, "the municipal authorities, in collaboration with property owners, set about widening some of the narrower streets in the Settlement and opening lanes to connect the main streets. To what extent this was carried out I do not know, but it was a very worthy project, as the streets of Yokohama were designed for rikishas [*sic*], not automobiles; and in the Japanese city the need for wider thoroughfares was even more pressing."[11] Thus even as Poole lamented the demise of "his" Yokohama, he could appreciate the intent of local authorities to build a more modern and livable city.

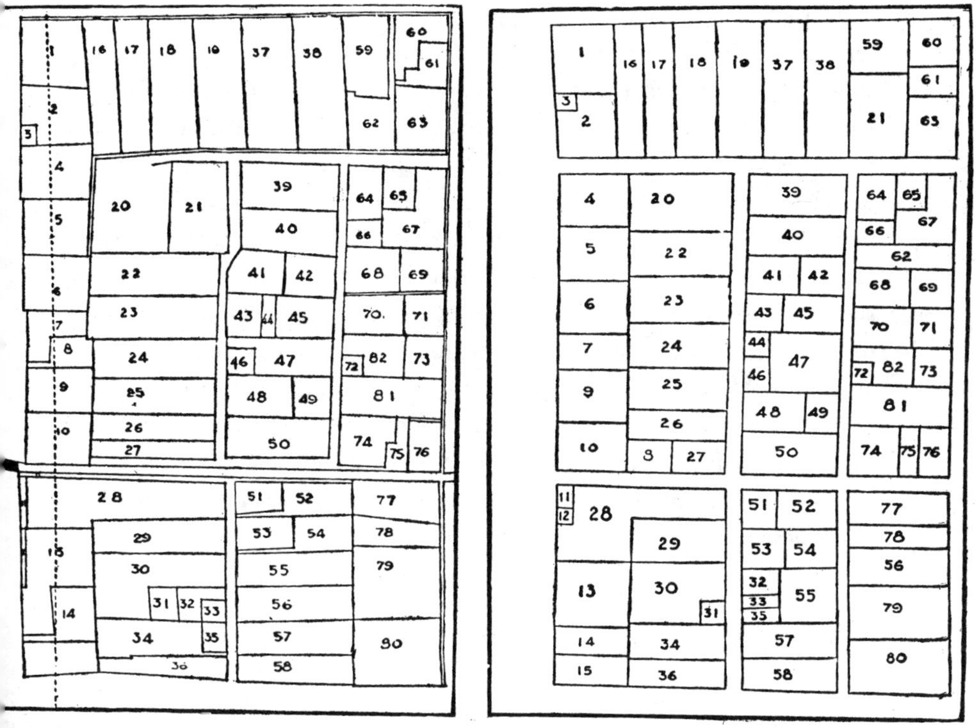

FIGURE 8.5. Before (left) and after (right) maps of a typical district in Tokyo. Following the earthquake, lots and streets became more regular, with wider streets and no more blind alleyways. (Page 269 from K. Sawada and Charles A. Beard, "Reconstruction in Tokyo," [March 1925])

Municipal authorities in Yokohama also took far-sighted steps to ensure their city's future as a commercial center. Faced with the prospect of losing the city's valuable silk trade, city leaders offered to reconstruct destroyed buildings, rent them to silk companies at 10 percent of the cost, and turn the buildings over to the companies after ten years. As Poole describes it, "It was a master-stroke. Up went the premises, back went the silk firms, and the valuable trade became once more a virtual monopoly of the northern port."[12]

Local authorities also took steps to ensure that ruined buildings along the harbor would be cleared in a timely fashion, converting the cleared building sites into parks. Over time new buildings began to appear, but the writing also appeared on the wall: Tokyo, not Yokohama, was destined to be the country's commercial center and, therefore, the focus of the revitalization efforts.

In Tokyo and elsewhere the earthquake helped modernize commerce at its most fundamental level: shopping. Prior to the earthquake, customer service was highly personal and time-consuming: customers removed their shoes, sat on a straw mat, and waiting for assistants to retrieve desired goods. Larger shops were at a disadvantage, given this custom, because of their inevitably slower service. A great many small shops were razed by the earthquake; some were rebuilt and back in business very quickly, but other businesses seized the opportunity to build Western-style department stores. Customers were free to wander around as they wished, and to keep their shoes on. New stores sprang up next to railroad stations, fueling growth of outlying districts; stores added amenities to attract customers, including restaurants. Since much of the shopping was done by women, women also did much of the eating at these early forerunners of food-court restaurants. This might not sound like an observation that bears mention, but prior to this time it had been considered improper for women to be seen eating outside of the home.

FIGURE 8.6. The main business section of a new and improved Yokohama, following reconstruction. (From tourist booklet, "Yokohama," printed by Ohkawa Printing Company, Yokohama, Japan [publication date unknown])

Playing the "what if" game, one could make the argument that the forces that led to Japan's involvement in World War II were not shaped in a serious way by the Kanto earthquake, but the post-earthquake modernization efforts set the stage for Japan's explosive postwar recovery, which in turn set the stage for the leading role that Japan now plays on the global stage. What if the Kanto earthquake had never happened? The question remains unanswerable, but this answer is at least as plausible as any other.

It is much easier to assess the impact of and subsequent recovery from the Kanto earthquake over the short term. In the immediate aftermath of the temblor, a large swath of coastal Japan found itself in utter disarray: no electricity, water, communications, transportation . . . a million people without food or shelter . . . an endless, heart-wrenching parade of survivors searching for the missing. A Provisional Relief Board toiled day and night for days after the earthquake to meet the basic needs of Japan's citizenry. Martial law was declared on September 2 and remained in effect until mid-October. With usual communication lines out of service, 400 carrier pigeons were dispatched with messages in the week following the earthquake.

In many ways, the hardest-hit areas came back to life remarkably quickly, in Tokyo especially. By September 4 electricity had been partly restored in Tokyo; telephone service was largely restored a day later. By September 6 some trams were running, and the post office began carrying some mail. Bank service began to be available to handle important transactions on September 10.

Not all services were restored overnight, of course: not until October 20 were water and gas services restored throughout the city. Seven weeks is a long time for a big city to be without such basic services, and the conditions took their toll. By the end of November, over 2,500 cases of dysentery had been documented, along with thousands of cases of other serious communicable diseases. More than 1,300 people died as a result of illnesses that the earthquake had unleashed.

By some accounts recovery was painfully slow, with foreign investors reluctant to venture back onto such clearly unstable soil. By the spring of 1929, however, the last of the ruined buildings were cleared away in Tokyo. In 1930 a festival was held to celebrate Tokyo's complete recovery.

From the vantage point of the 21st century one can now look back and consider the world's impressions of Japan's recovery, impressions that were perhaps inevitably colored by different perceptual filters over the years. The October 1923 issue of *National Geographic* magazine included a long article, "The Empire of the Risen Sun," that was obviously planned and written before the earthquake. This article, by William Elliot Griffis, spoke in glowing terms of a country and its people. "The acorn of 1870," Griffis wrote, ""has become the oak of 1923; but it was planted in a soil made of the enriched mold of a thousand years of culture."[13] The editor added a preface to this article, speaking of an "industrious and resourceful people" and noting that "The same qualities and characteristics which Dr. Griffis . . . sets forth as responsible for the rise and development of the Japanese will enable them to build a new and greater capital from the ruins of the old. The fortitude of the island kingdom of the East in the face of unprecedented disaster has commanded the admiration of the entire Western World."[14]

By 1943, just twenty years later, Western views of Japan and its people of course could not have been more different. The August 16, 1943, issue of *Life* magazine featured a cover story, "How Strong Is Japan?" And a two-page article in the same issue includes sickening photographs from 1923: fire victims at the Honjo military garment depot, and waterfront carnage caused by the tsunami. That these photographs are described as "hitherto unpublished" speaks volumes about the nature of wartime versus peacetime sensibilities. So, too, does the article itself, which concludes, "The dogged, ant-swarming energy of the Japs rebuilding their shattered country gave Occidentals their first real idea of the frightening persistence of this strange little people."[15] As deeply as these words offend modern ears, they provide yet another testimony to Japan's recovery. Even a wartime writer speaking of an enemy nation had to acknowledge the remarkable nature of the rebound, leaving himself only room to portray the recovery in darkly ominous rather than admirable terms.

The 1923 Kanto earthquake was one of the most devastating direct hits on an urban population center in the 20th century. Without question the earthquake caused devastating losses of both life and property. Tokyo and the surrounding cities and villages did not rebound right away, but they did rebound within no more than a handful of years. In a small tourist booklet published in the 1930s, the mayor of Yokohama presented a photographic tour of the city's many new, modern structures:

New Yokohama as shown in this book was rebuilt by the people of Yokohama with their undaunted courage and generous aids given at home and abroad. The reconstructed Yokohama is the best planned city in Japan with her extended area and greater population than before the earthquake.

> The future of Yokohama as gateway to the Orient and as one of the leading business and industrial centers of Japan, is bright and promising, for her foreign trade and industry are not only recovering but they are also developing more and more according to the growth of the trans Pacific commerce and shipping.[16]

City officials can be expected to paint only an optimistic face on the portrait of their economic condition. But photographs do not lie, and the ones in this small book show building after building, shiny and modern and new.

This apparent success story notwithstanding, one cannot help but ponder the question of what will happen when (not if, *when*) Tokyo is struck by another earthquake as large as that of 1923. In 1989 the Tokai Bank estimated that a repeat of the Kanto earthquake would cause $650 billion of damage in the Tokyo area alone; a 1994 study by a California engineering firm estimated total losses of $800 billion to $1.2 trillion. Such estimates will of course depend critically on the valuation of Tokyo real estate, which has fallen since the heady days of the late 20th century and may either fall further or rise in the future. Still, according to some projections, losses from the next great earthquake disaster in Tokyo could be large enough to trigger a global economic crisis.

Without question, the Japan of the early 21st century is better prepared for future earthquakes than the Japan of the early 20th century, yet far more lives and structures are now at risk. At M6.9, the 1995 Kobe, Japan, earthquake was humble compared with great subduction zone earthquakes such as that of 1923, yet this temblor claimed over 5,000 lives and caused tens of billions of dollars in property damage. The much larger M8.3 "Takachi-oki" earthquake of September 26, 2003, caused far less damage and zero fatalities. But, as is the case with real estate prices in general, three key factors control the likelihood of damage during a large earthquake: location, location, location. The 2003 temblor was centered about 60 kilometers offshore; a repeat of the great Kanto earthquake would strike the heart of urban Japan—a direct hit on Tokyo, whose population is now ten times higher than it was in 1923.

As tragically illustrated by the M6.6 earthquake at Bam, Iran, in 2003, the greatest vulnerabilities to earthquake damage are within developing nations, in particular those with a tradition of adobe or masonry construction. Among anticipated future earthquakes in industrialized nations, however, perhaps none looms more menacing than the inevitable repeat of the 1923 Kanto event. A consideration of history suggests that the next great Tokyo/Kanto earthquake will not be the death knell of modern Japan. After tens of thousands of years, it seems unlikely that the human and societal capacity for rebound has suddenly reached its limit. However, a consideration of history—in this case as well as others that later chapters will discuss—clearly reveals that, the capacity for rebound notwithstanding, societies must continue to take steps to reduce future earthquake losses. Many such steps have been taken in Japan, which has enacted strict building codes and spends far more per capita on earthquake monitoring and research than does the United States. With a temblor like that of 1923 in its not-too-distant past, Japan understands all too well that the fact that such losses are *survivable* is irrelevant: the point is that they are *avoidable*.

9

Hazards of the Caribbean

The scene in this palm-grove was not unlike an old-time camp-meeting. The sojourners in both tents devoted their time principally to religious exercises, of which singing formed the greater part. At times these tents would be giving forth volumes of music and praise that made the very welkin ring; but in a day or two it came to be the custom to alternate, one listening while the other sang, until the superiority of the negro music was acknowledged.

—Louis Housel, "An Earthquake Experience,"
Scribner's Monthly 15, no 5 (1878): 671

The Caribbean is a place of romance. Idyllic beaches, buoyant cultures, lush tropical flora; even the Caribbean pirates of yore often find themselves romanticized in modern eyes, and on modern movie screens.

Yet it requires barely a moment's reflection to appreciate the enormous resilience that must exist in a place that is so routinely battered by storms of enormous ferocity. News stories tend to focus on large storms that reach the United States, but many large hurricanes arrive in the United States by way of the Caribbean. Before it slammed into South Carolina in 1989, Hurricane Hugo brushed the Caribbean islands, skimming Puerto Rico and devastating many small islands to its east. Other hurricanes have hit the islands more directly. These include Inez, which claimed some 1,500 lives in 1966, and the powerful Luis, which caused $2.5 billion in property damage and 17 deaths when it pummeled the Leeward Islands and parts of Puerto Rico and the Virgin Islands in 1995. Hurricanes also figure prominently in the pre-20th-century history of the Caribbean—storms that had no names, the sometimes lethal fury of which arrived unheralded by modern forecasts.

Most people know that the Caribbean is hurricane country; probably few realize that it is earthquake country as well. After all, the western edge of North America is the active plate boundary; earthquakes occur in the more staid midcontinent and Atlantic seaboard, but far less commonly. What can be overlooked, however, is North America's *other* active plate boundary. To understand the general framework of this other boundary, it is useful to return briefly to basic tenets of plate tectonics theory. As discussed in earlier chapters, the eastern edge of North America is known as a passive margin. Because the North American continent is not moving relative to the adjacent Atlantic oceanic crust, in plate tectonics terms, scientists do not differentiate between the North American continent and the western half of the Atlantic ocean. That is, by defining plates as parcels of crust that move as more or less coherent blocks, these two seemingly distinct terrains are part of a single North American plate. The eastern edge of the North American plate is marked by a submarine volcanic range whose fires cut Iceland in half.

But consider the following question: The North American plate must have a southern edge as well as side edges—where is it? Answer: It cuts across southernmost Mexico through Guatemala and heads more or less due east into the Caribbean. To the south of the eastern part of this plate boundary lies the relatively small Caribbean plate. Considering the principles outlined above, it is clear that the Caribbean plate would not have been defined in the first place were its boundaries not active. Especially when considered on a square-mile basis, the Caribbean plate is an interesting and complicated piece of the plate tectonics puzzle. Just as the dynamic, multicultural culture of western North America can arguably be considered to mirror the geology and tectonics of the region, so one can draw parallels between the tectonics of the Caribbean plate and the especially vibrant and multihued cultures of its islands. In this chapter we embark on a short detour, largely leaving aside discussion of earthquakes' impact to focus on an overview of a region whose unique and dynamic geology is to a large extent unknown outside the halls of science. The issue of tsunami risk in particular emerges as relevant in the wake of the 2004 Sumatra disaster.

The most distressing feature of the Caribbean plate to a seismologist is that, of the many millions who now populate this idyllic region, only a few hundred people live at its core, far from the earthquakes that mark its edges. Some 99 percent of the peoples of the Caribbean live perilously between active faults and active volcanoes on its slowly moving plate boundaries. Two tiny islands alone protrude above the central inactive plate, which like North America is made from ancient continental rocks: the islands of Andreas in the west and the island of Aves in the east. Aves Island, which in the 1800s was a few kilometers across, all but disappeared when its protective coating of guano was mined as fertilizer later that century. It is now a sandbank the size of a football field that hosts a million birds (as its name implies) and an elevated lighthouse staffed by the Venezuelan navy.

Of particular relevance for this chapter is the active plate boundary along which the Caribbean islands are strung like pearls. In a broad-brush sense the Atlantic seafloor moves westward away from the Mid-Atlantic Ridge, diving under the Caribbean plate somewhat obliquely, raising the island of Barbados like a layer cake, and raising hell in the form of the numerous volcanoes, starting with the submarine volcano Kick'm Jenny (named for the violent but invisible explosions that rattled the timbers of passing ships) near Grenada in the south and terminating near the demonically explosive volcanic domes of Montserrat and Martinique in the north. The collision of the northeast corner of the Caribbean plate with the Atlantic seafloor is somewhat more complex. The eastward motion of the corner of the plate over the descending Atlantic seafloor has raised the granite and corals of the Virgin Islands, and with them the seafloor, to levels that have been the doom of a thousand ships. A complexity arises in the form of the Bahama platform, a tectonic block now embedded in the North American plate and surfaced by limestones and corals, that 60 million years ago bore the brunt of a head-on collision with the Caribbean plate. The Bahama platform continues to prevent Hispaniola from moving eastward along with the rest of the Caribbean plate, but Puerto Rico, unhindered by the platform, has splintered off from Hispaniola along faults that run obliquely between the islands.

Thus Puerto Rico's most important faults are found offshore, principally the thrust fault along the subduction zone interface to the north of the island, and to a lesser extent along the Mona Passage between Puerto Rico and Hispaniola. Where oceanic crust subducts, there is typically a trench (figure 9.1), and the Puerto Rico Trench holds the distinction of being the deepest part of the Atlantic Ocean. One might expect oceans to be deepest

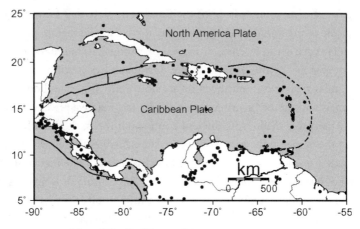

FIGURE 9.1. Map of the Caribbean plate.

in the middle, but along this trench, tucked away in an inconspicuous corner, the sea bottom is as low as 8,400 meters (over 5 miles!) below sea level. The reasons for this are not entirely understood, but geophysicist Uri ten Brink recently proposed an intriguing model. According to ten Brink's interpretation, the curvature of the subduction zone has caused a tear in the downward-sloping crust, creating a particularly narrow piece of subducting slab that extends westward the length of Puerto Rico, starting from offshore of the northeast edge of the island. Narrow slabs are known to experience a phenomenon known as rollback, whereby the descending plate, instead of diving at an angle into the earth's mantle, sinks vertically, forming a steeply dipping subduction zone.

In regard to hazards, the salient feature of the trench is that it lies offshore by some 50 kilometers. Thus even major subduction zone earthquakes will occur at some distance from dry land, their impact diminished by the distance over which earthquake waves must travel before they reach people or structures. The fault, or rift, zone to the west of Puerto Rico also lies offshore. The impact of temblors on these fault zones may be blunted by distance, but it is scarcely obliterated. As we will discuss shortly, offshore earthquakes of appreciable size—typically magnitude 7 to 7.5—have struck Puerto Rico in historic times, producing both damaging shaking and, in some cases, tsunami. Both direct eyewitness accounts and sophisticated computer modeling reveal that tsunami can reach heights as great as 10 meters along the northern coast of Puerto Rico, high enough to wreak havoc in low-lying coastal areas.

To the south of Puerto Rico the seafloor topography, or bathymetry, reveals another trench, but modern surveys with GPS instruments suggest that this feature is, at least for now, dead as a doornail. That is, while subduction must have occurred to the south of Puerto Rico at some point in the past, it does not appear to be active today. The question is, how long has it been dead? The answer is critical for hazards purposes: Will it stay dead? Some geologists view the present-day quiet of the southern subduction zone as very temporary. Viewing historic and prehistoric earthquake patterns worldwide, geologist James Dolan sees a tendency for such dueling fault systems to switch on and off in turn, each staying active for perhaps several hundred years.

For the purposes of earthquake hazard assessment, it is critical to understand not only the major offshore plate boundary faults but also the active faults within Puerto Rico itself. Although they are not part of the plate boundary proper, faults within Puerto Rico pose a substantial concern for two reasons. First and foremost, these faults are closer to the population centers, and could well be especially damaging even if of relatively modest size. Even in earthquake-prepared California, the relatively modest M6.7 Northridge earthquake of 1994 proved to be the most expensive earthquake of the 20th century because it caused high ground motions in the densely populated Los Angeles metropolitan area.

Moreover, Puerto Rico's inland faults are very poorly understood; only the smallest handful have even been identified, and none of these have been studied in detail. Fault investigation within Puerto Rico has proven especially challenging for several reasons. The commonwealth has received relatively little attention over the years from the scientific community, a reality that reflects political as well as geologic priorities. Although damaging earthquakes have struck Puerto Rico throughout its history, earthquake hazard has appeared to be relative low, and less of a concern than the hurricanes that more commonly batter the island. The terrain itself also complicates efforts to investigate faults in Puerto Rico; much of the island is heavily forested and subject to high rates of erosion. In the waning years of the 20th century, however, geologists began to identify and investigate active faults on the island.

In this chapter we depart somewhat from the theme of rebound to focus instead on a geologic tour of North America's other plate boundary. To begin this tour, we first take a step back in time to consider some of the important historic earthquakes in and around Puerto Rico.

Earthquake History

For the seismologist interested in characterizing the earthquake history of
Puerto Rico, the fairly simple history of the island is a cause for celebration.
In contrast to the more turbulent histories of other Caribbean islands, Puerto
Rico remained a Spanish possession from its discovery by Christopher Colum-
bus in 1493 until its transfer to U.S. hands in 1898, following the Spanish-
American War. Archival accounts of historic earthquakes are therefore relatively easy to find and interpret, for one generally needs to deal with only a
single country and records in no more than two languages. The most complete search of archival sources, including early official correspondence, was
done by seismologist William McCann. Focusing exclusively on original
sources, McCann was able both to better elucidate some known historic temblors and to disprove the existence of others that had found their way into
the catalog. This work culminated in a list of the most significant historic
earthquakes in Puerto Rico.

Archival sources from Puerto Rico's earliest days as a Spanish possession
are few and far between. They include letters sent from Puerto Rico to the
king of Spain and the pope. Two Catholic dioceses were established on Puerto
Rico in 1511; communications from these dioceses refer vaguely to earthquakes that must have occurred between 1493 and 1511, possibly around 1502.
The island had few inhabitants during these early years, however, and extant
accounts are both limited and sorely lacking in detail. In such a situation it
is impossible even to be sure that any earthquakes occurred; there is precedent for large storms to eventually give rise to tales of nonexistent large
earthquakes.

The first clearly established earthquake in Puerto Rico occurred in 1670.
The San German district in southwestern Puerto Rico was severely affected,
but our understanding of this earthquake and its effects remains sketchy.

One of the most important earthquakes in Puerto Rico's history struck
around midday on May 2, 1787. Relatively detailed accounts describe damage
to military fortifications along much of the island's north shore. Accounts
also describe widespread damage to houses, although, in keeping with expectations for an offshore event, the mistreatment did not generally extend
as far as serious damage.

In 1867 Puerto Rico experienced the first major earthquake for which we
have any semblance of modern scientific understanding. Reverberations
from this temblor echoed throughout Puerto Rico and the Virgin Islands,

with damage concentrated in eastern Puerto Rico. Witnesses described two distinct shocks approximately ten minutes apart. Two tsunami waves came ashore at Charlotte Amalie on St. Thomas. Accounts describe walls of water 4.5–6.1 meters high, waves that destroyed all of the buildings along the low-lying waterfront. At Fredrikstad on St. Croix the waves were even fiercer, reaching heights of 7.6 meters. At this location the first wave came ashore immediately on the heels of the shaking, suggesting that the tsunami originated close to Fredrikstad—possibly from a local seafloor slump in this location rather than actual motion along the fault.

A detailed and dramatic account of the 1867 tsunami was penned by Louis Housel, at the time a freshly minted graduate of Annapolis assigned to the U.S.S. *Monongahela*, described by Housel as "one of Uncle Sam's most unseaworthy double-enders."[1] Once again, a detailed and eloquent firsthand account has a singular power to carry us back in time—to give us a taste not only of experience almost beyond imagination, but also of the depths of human resilience.

The (much-maligned) *Monongahela* arrived in Fredrikstad on November 17, 1867, when the weather was seasonally warm and unremarkable. At about 3:00 P.M. on November 18, all hell broke loose. The sailors' attention was first captured by vibrations of the ship: "Our vessel began to quiver and rock as if a mighty giant had laid hold of her and was trying to loosen every timber in her frame."[2] The vibrations, reportedly accompanied by a buzzing noise, led the men to suspect a problem with the ship, perhaps a fire in the boilers.

"It's an *earthquake*, sir; look ashore!" came the shout from one sailor on the bow, pointing to the shore, where the town was engulfed in a dusty haze and people ran frantically to and fro. Part of the stone tower of the English church—an especially large structure visible from the ship—had collapsed.

Some five minutes after the vibrations subsided, a new sound arose in their place, described as a "peculiar grating noise." Peering over the bow, Housel discovered the origin of this sound: a full ten fathoms (20 meters) of the anchor chain was suddenly out of the water and pulled taut. An order was given to break the chain, but the enormous force of the waves succeeded ahead of the sailors' valiant efforts: "A tremendous jerk and the heavy fourteen-inch bolt riveted in solid oak beam was torn out and the last links connecting the vessel to the anchor went flourishing and wriggling overboard with the rest."[3]

Although tsunami are generally known as walls of advancing water, the prelude to many tsunami is the departure of the sea toward the horizon, exposing flailing fish to tempt unwary villagers, who are drowned by the returning ocean wave. To sailors anchored offshore of Fredrikstad in 1867, the massive retreat of the ocean from the shore, amid a sea suddenly chaotic and terrible, revealed a reef off the northern point of the island where before there had been several fathoms of water (figure 9.2). The returning current carried the ship toward shore, where eventually she smashed a storehouse and broke a row of trees. Before the crew of the *Monongahela* could respond, the ship was carried some 500 meters back out to sea.

Although the immediate danger had abated, an even more frightening one literally rose to take its place.

> That immense body of water which had covered the bay and part of the town was re-forming with the whole Atlantic Ocean as an ally, for a tremendous charge upon us and the shore.
>
> This was the supreme moment of the catastrophe. As far as the eye could reach to the north and to the south was a high threatening wall of green water. It seemed to pause for a moment as if marshaling its strength, and then on it came in a majestic unbroken column, more awe-inspiring than an army with banners. The suspense was terrible![4]

The crew of the *Monongahela* watched the wave approach, certain they were doomed to perish. Astonishingly, "not a drop of water reached [the] decks." The initial approach of the massive wave tipped the boat over on the starboard beam ends; as the ship righted itself, it was "buoyed to the crest of the wave and carried broadside to the shore, finally landing on the edge of the street in a cradle of rocks that seemed prepared for her reception. Here she rested with her decks inclined at an angle of fifteen degrees."[5]

Although badly battered by the tumultuous experience, thus did the "unseaworthy" U.S.S. *Monongahela* ride out the 1867 tsunami against all odds. Fighting their way to dry land and safety, the crew were not so distraught that they failed to appreciate a few notes of comedy in the scene. "This water bore on its surface all manner of *débris*, which it had gathered from the yards and houses in its course,—chairs, cradles, bedsteads, broken fences and doors, together with flocks of ducks and geese quacking and gabbing, utterly bewildered by the sudden rise of their natural element."[6]

And one can scarcely fault the ducks and geese in their response. On shore the crew found local residents as badly battered as local structures. The crew crafted large tents from the ship's awning, one for whites and one for blacks.

FIGURE 9.2. (Above) Sketch of the *Monongahela* during the 1867 tsunami. This tsunami is reported to have exposed the seafloor for a distance of half a mile, indicating an outgoing tsunami wave. (Like ocean waves, tsunami waves retreat and advance.) The incoming crest then followed, as depicted here in the distance. (Left) Photograph of the U.S.S. *Monongahela* on dry land. (Sketch from Louis V. Housel, "An Earthquake Experience," *Scribner's Monthly* 15, no. 5 [March 1878]; photograph courtesy of the U.S. Navy)

The scene in this palm-grove was not unlike an old-time camp-meeting. The sojourners in both tents devoted their time principally to religious exercises, of which singing formed the greater part. At times these tents would be giving forth volumes of music and praise that made the very welkin ring; but in a day or two it came to be the custom to alternate, one listening while the other sang, until the superiority of the negro music was acknowledged.[7]

As the aftershocks diminished in frequency, the music took on a sometimes joyous quality. "On one occasion they were singing with great gusto, 'I wish I were in Dixie,' when whir-r-r-er came a tremendous vibration, which hushed every voice in an instant."[8] Following a flurry of prayer, the music resumed on a different note, "On Jordan's stormy banks I stand."

Historic accounts sometimes come as an affront to modern sensibilities, since they inevitably reflect the situation of their day—not ours. But if Housel was inclined to portray Fredrikstad's black population as especially unsettled (so to speak) by the tumultuous events, his closing comments make clear the extent of the unease felt by all.

All, however, seemed to suffer acutely from anxiety and nervousness. There is nothing, I believe, so trying to a healthy nervous system as a succession of earthquakes. To a landsman a gale at sea has untold terrors; yet the tossings of his bark can be accounted for: the wind and waves are there, and the result may be anticipated. But in an earthquake all these factors are wanting; the cause is mysterious and unknown; the result anticipated is destruction in some form, and the tension of the nerves is most wearing.[9]

Firsthand accounts of devastating earthquakes and tsunami can paint a scene with singular effectiveness and drama. To understand more fully the effects of the 1787 and 1867 temblors, it might be helpful to set the stage by traveling to the graceful, well-worn cobblestone streets of old San Juan that evoke thoughts of old Spain. Following these streets to the northwest corner of the district, one arrives at El Morro, the largest colonial fortification in the Caribbean. Constructed by the Spanish in the 16th century, El Morro has 6-meter-thick walls that rise 42 meters above sea level. Following centuries of service as a fortress, El Morro was converted to a museum, but it is much more than that—it has become the national symbol of Puerto Rico, a National

Historic Site administered by the National Park Service. On even the hottest, most languid days on the island, one can find temporary respite in the climate-buffered deeper rooms of El Morro and in the cooling sea breezes sailing above its upper levels.

The walls of El Morro stand tall and seemingly invulnerable to anything nature might have to offer, but the earthquake and subsequent tsunami of 1787 caused considerable damage to what was then the castle of San Felipe del Morro as well as to the nearby castle of San Cristobal. Cisterns, walls, and guard houses were broken. Elsewhere on the island the Arecibo church was damaged severely, as were churches in other towns, including Mayaguez. Damage from the 1867 quake was concentrated in the eastern part of the island, since the temblor was located in the Anegada Passage between Puerto Rico and the Virgin Islands.

As discussed in earlier chapters, the year 1900, give or take a few years, marks a critical watershed in earthquake studies. Major earthquakes occurring after this date were typically recorded by at least a few seismometers around the globe. Thus we have no instrumental recordings of the 1867 temblor but fairly good data with which to investigate an earthquake that struck off the northwest corner of the island in 1918. Analyzing early seismometer recordings, seismologists Javier Pacheco and Lynn Sykes obtained a magnitude of 7.5 for this temblor, along with—at least relative to earlier Puerto Rican events—a good estimate of its location. Comparing the location and sense of faulting of this earthquake with our modern geologic understanding of the Caribbean, it appears that the earthquake occurred in the Mona Canyon, the rift across which Puerto Rico pulls away from Hispaniola.

Because the 1918 temblor occurred in relatively recent times and after the advent of photography, we have good documentation of the damage it caused throughout the island. The earthquake struck without warning on October 11, 1918. Witnesses described two severe shocks separated by perhaps two to three minutes, the first being the more severe.

In Mayaguez the wooden columns supporting a porch were apparently thrown into the air: when the shaking ceased a shoe was found caught between the base of one column and the floor of the porch. Relatively severe effects occurred throughout Mayaguez even though it was farther from the earthquake origin than other towns that escaped with less damage. As is so often the case, this is largely attributable to the nature of the subsurface geology upon which Mayaguez is built: sedimentary ground, in many places saturated by water. In 1918 the effects were dramatic. Substantial concrete and reinforced concrete buildings had only light damage, but some poorly built concrete buildings fared much worse. Several lives were lost in the collapse of "La Habanera," a large two-story cigar factory in a lower part of the city. Even when walls did not collapse, the shaking often took a toll on plaster and weak architectural elements.

The 1918 earthquake generated a tsunami that came ashore along the west coast of the island and caused 114 deaths.

For both the older historic earthquakes in Puerto Rico and those during the early 20th century, earth scientists have two principal investigative tools in our bag of tricks—beyond the analysis of whatever seismometer recordings exist. The first of these, alluded to above, is analysis of damage accounts. The second is investigation of preserved geologic evidence, particularly of liquefaction. Direct geologic investigation of offshore faults is often impossible, but with ingenuity and painstaking care, scientists can unearth (so to speak) evidence of secondary effects caused by an earthquake.

As is the case in many other regions, liquefaction features have provided arguably the most important data with which the historic earthquakes of Puerto Rico can be investigated. We know that liquefaction occurred during large historic events on the island; surviving accounts say so. Given the geologic setting of the island, liquefaction during large earthquakes certainly cannot be considered a surprise in view of scientists' experience in other regions.

Unfortunately for geologists, Puerto Rico's geology both giveth and it taketh away. The saturated coastal and river valley soils create both conducive conditions for liquefaction to occur and lush vegetation that greatly complicates efforts to find and investigate these features. In a less tropical climate such as New Madrid, ample evidence of 200-year-old sand blows can readily be seen in even modern air photographs and ground surveys. In Puerto Rico, surface evidence of sand blows is quickly erased. Surface reconnaissance therefore proves useless over most of the island, but, undaunted, intrepid geologists capitalize on the windows of opportunity provided by the island's rivers. Along these corridors flowing water excises soft-soil riverbanks, providing linear portals into the island's subsurface.

In areas with extensive soil development, riverbanks will of course be dynamic features. At any given time they will reveal a single cross section, or slice, along the bank. Following storms or continued erosion, a riverbank will be recarved and reshaped, revealing different slices of terrain. Imagine a yellow cake made with raisins; possibly an unappealing concoction, but don't think about that. Instead, imagine slicing the cake in half. Viewing one of the halves crossways, one would see a certain number and distribution of raisins. Now imagine slicing an inch or so off one half; discarding the thin slice and viewing the remaining cake, one would see different raisins. If the cake batter was well filled with raisins, most slices would reveal some raisins, although always a different number and pattern.

Now imagine making progressive slices into a chocolate cake baked with chocolate chips. The situation would be the same in principle as for the cake with raisins, but different in practice: chocolate chips look a lot like chocolate cake, and so will be much more difficult to discern. Chocolate chips also have more of a tendency than raisins to smear themselves out, and thereby lose their distinct edges. Our chocolate-on-chocolate cake starts to represent a more apt analogue for liquefaction features in a geologic setting such as that found in much of coastal Puerto Rico. Along the riverbanks that flow toward the island's northern coast, geologists search for evidence of sand dikes: vertical conduits through which sand has traveled upward to the surface during liquefaction, and in which sand still remains. To find such features, one must distinguish one type of dirt from another, the difference being that the dirt within sand dikes is characteristically sandy while the surrounding dirt has a higher mud content. The width of sand dikes varies considerably, from tiny features as small as a centimeter in width to as much as several meters in the most extreme cases.

Some types of dirt are more easily distinguished than others, however, and along Puerto Rico's riverbanks some liquefaction features can be spotted by the trained—and even the untrained—eye. Geologist Martitia Tuttle and her colleagues have combed many miles of riverbank in Puerto Rico and have identified an ever-growing collection of liquefaction features. The largest concentration of features has been found along the Rio Grande de Anasco, but features have been found along other rivers as well, including the Rio Culebrinas, which drains northwestern Puerto. Some of these liquefaction features, particularly dikes in which the sand has oxidized to a characteristic umber, are indeed recognizable to the untrained eye (figure 9.3).

The age of sand blows and dikes can be estimated if one can obtain carbon-14 dates of surrounding soils. For example, if a sand dike cuts through material that is 500 years old and is capped by undisturbed soils 300 years old, the dike must have formed between 1500 and 1700. The development of such results involves substantial expertise with soils, painstaking fieldwork, and often a little luck. With at least as much of the first two as the

FIGURE 9.3. Dark sand indicates dikes that were formed by liquefaction during a large historic earthquake in Puerto Rico. By finding such features and determining their ages, geologists are able to learn more about the earthquake history of the Caribbean. (Susan E. Hough)

third, Tuttle and her colleagues have identified and dated both sand dikes and sand blows, the dates of which can be compared with the dates of known historic earthquakes. The dates of some features are consistent with known historic events, including the 1670 and 1918 temblors, and an M7.5 that occurred some 60 kilometers offshore in 1943. A number of features appear to predate the historic record, with estimates of their having occurred between 1300 and 1508.

In some regions worldwide, paleoliquefaction investigations have allowed scientists to identify the approximate times and magnitudes of prehistoric earthquakes, earthquakes whose existence was not previously known. In a situation such as one finds in Puerto Rico, where liquefaction features primarily correspond to known historic events, this type of investigation is still tremendously useful. It often improves estimates of magnitudes of historic earthquakes; it also can tell us quite a bit about ground shaking during the events, since a certain level of shaking is required to cause liquefaction.

Traveling up the Rio Culebrinas on a modest motorboat, you can almost imagine yourself in the Amazon, the air sultry and the riverbanks lush with vegetation. Upon closer inspection, the scene is definitely Puerto Rican, however: the semiwild but mild-mannered dogs that scamper alongside, and sometimes in, the water; the balmy breezes and indolent pulse of the day; the nonindigenous but fat and contented iguana sunbathing on a branch just above the water. Rivers such as this provide the portal through which geologists can view and explore Puerto Rico's geologic past.

Another tool in the geologist's bag of tricks is the study of uplifted marine terraces: shorelines characterized by features including, commonly, a broad and flat expanse and the presence of corals that have clearly been lifted up out of the intertidal zone in which they grow (figure 9.4). Shorelines can become uplifted (beached, as it were) if sea level is lowered or the land is raised. Scientists know the time at which sea level was most recently significantly higher than it is today, the high sea level stand: 120,000 years ago. Where one comes across an uplifted shoreline, one must first consider whether it represents this sort of global change in sea level or a local change in land level.

The identification of marine terraces is relatively straightforward, given their characteristic morphology and, typically, the presence of corals and shells on land that is now high and dry. What is much more difficult is the detailed analysis required to unravel the history of a marine terrace; typically one uses dating techniques to establish how long it has been since the corals were alive. Along the southern coast of Puerto Rico, near the town of Salinas, a conspicuous terrace has been identified. The terrace can be traced in a 1937 aerial photograph that predates later modification of the landscape. Several lines of evidence suggest local uplift of the land during a large earthquake, but detailed geologic investigations have been slow in coming.

Elsewhere on the island, geologists have begun to make more progress in their quest to locate faults and determine their activity over the recent geologic past. The overall topography of Puerto Rico is clearly shaped by two large fault zones that run nearly stem-to-stern across the island: the Great Northern fault and the Great Southern fault. The surface expression of the latter is particularly clear: a narrow ribbon diagonally across the island from the midsouthern shore to the midwestern shore. While both faults clearly expe-

FIGURE 9.4. Large subduction zone earthquakes can cause the entire coast to be elevated substantially, forming stair-step features known as marine terraces. The terrace shown here is in Alaska. (U.S. Geological Survey)

rienced substantial motion at some point in the geologic past, scientists have found no evidence that they remain active today.

The first recognized and investigated active fault on Puerto Rico lies within the Lajas Valley in the southwestern portion of the island. The Lajas Valley is an east–west-trending, 30-kilometer valley bounded on both sides by mountains. The overall morphology of the valley and local drainage patterns suggest the presence of an active fault; again, early aerial photos (1936) provided critical evidence of fault features, in this case a scarp in the southwest part of the valley (figure 9.5). Geologist Carol Prentice and her colleagues zeroed in on this especially abrupt linear feature in the landscape, and in the year 2000 were able to conduct a one-day reconnaissance study. This study, which involved excavation of a 2-meter-deep trench, revealed evidence of recent movement on a nearly vertical fault. Prentice and her colleagues returned to the site in 2003 to launch a more substantive investigation. This work con-

FIGURE 9.5. Most of the important faults in the Caribbean are offshore, but geologists have begun to identify faults onshore as well. The modest-looking ridge in this photograph is one such fault, investigated by the geologist Carol Prentice and her colleagues. (Susan E. Hough)

firmed the initial conclusion that within the past 7,500 years, this fault has experienced at least two earthquakes large enough to cause surface rupture.

Considering the topography and drainage patterns elsewhere on Puerto Rico, one can identify other likely suspects. One of these is along the western coast of the island not far west of Mayaguez, where another conspicuously linear valley trends in a southeasterly direction inland from the coast. This feature appears to be the on-land continuation of an offshore fault zone that has been identified by marine surveys. If a fault does indeed run through this valley, it could pose a significant hazard to the nearby city of Mayaguez.

As geologists continue their efforts to investigate faults within the island of Puerto Rico, they remain mindful that such faults will almost certainly be less active than the major plate boundary faults offshore of the island. Yet as experience has shown us time and time again in other regions, damage caused by an earthquake depends critically on its proximity to structures. The building codes in Puerto Rico have been written largely on the basis of the expected sizes and effects of offshore earthquakes. Houses, typically of concrete construction, are built with rebar reinforcement. Not uncommon on the island are single-story homes with multiple rebar antennae above their roof lines, spidery testimony to their owners' optimism that a second story will some day rise around the reinforcements. If the building codes do their job, these houses will ride out future offshore earthquakes relatively unscathed.

But will houses and other structures withstand the kinds of earthquakes that might occur on Puerto Rico's inland faults? When we talk about "inland Puerto Rico faults," it is perhaps useful to keep scale in mind. Whereas faults in the midcontinent of North America are often a few thousand kilometers away from the nearest active plate boundary, no point on Puerto Rico is more than perhaps 150 kilometers from the active boundary between the North American and Caribbean plates. The faults within the island are therefore not midcontinent faults but what geologists usually term plate-boundary-related faults. Moderate-to-large earthquakes on such faults can be expected at a relatively healthy clip relative to that typical of true intraplate faults. Geologists have recognized this reality for some time, but only relatively recently have marshaled the resources necessary to understand this complex region—America's other plate boundary.

Faults Elsewhere in the Caribbean

This chapter has focused on faults and earthquakes within and adjacent to Puerto Rico but, as noted earlier, all of the Caribbean islands are strewn along the same active plate boundary. To the east of Puerto Rico the subduction zone continues, and poses a similar hazard to the neighboring Virgin Islands. Indeed, the 1867 tsunami is often known as the Virgin Islands tsunami.

To the west of Puerto Rico the plate boundary character changes, giving way to a primarily lateral boundary by the time one reaches Hispaniola. From a hazard standpoint, the different style of faulting is less important than another change: along Hispaniola the plate boundary comprises a number of distinct fault strands, one of which runs not offshore, but through the island itself. Geologist Paul Mann and his colleagues first identified the left-lateral Septentrional fault as a major on-land plate boundary fault that extends for several hundred kilometers across northern Hispaniola. Carol Prentice and her colleagues found and published the first geologic evidence for large prehistoric earthquakes on this fault. Later work by Paul Mann and others further elucidated the complex plate boundary, which includes both the on-land left-lateral fault and a major subduction zone just offshore. The latter represents a continuation of the subduction zone offshore of Puerto Rico. This westward extension poses a hazard to Hispaniola similar to that posed by the eastward segment to Puerto Rico, but the on-land Septentrional fault poses an arguably greater hazard by virtue of its proximity to population centers.

Considering the known large earthquakes in the historic catalog for Hispaniola, geologist James Dolan came to an interesting conclusion, which we alluded to earlier: the subduction zones to the north and south of Hispaniola appear to take turns being active. Between 1615 and 1860, eight moderate-to-large earthquakes appear to have occurred in southern Hispaniola, whereas since 1842, four large earthquakes have occurred in and offshore of the northern side of the island. Although the locations and magnitudes of the earlier events are uncertain, one can find precedent worldwide for this type of behavior: earthquake activity ping-ponging back and forth, on time scales of centuries, between nearby major fault systems. Such oscillatory behavior between pairs of fault systems was first identified by Nick Ambraseys, in Turkey, where the northern and eastern Anatolian faults alternately become active, and stay active, for centuries at a time. Why fault systems behave this way is a puzzle; *that* they behave this way has only recently been realized via painstaking compilation and investigation of earthquakes that predate the brief century of modern instrumental seismology.

Earth scientists continue to probe the mysteries of complex fault systems in part because they represent a scientifically challenging and interesting problem, and in part because of the obviously enormous implications for earthquake hazard. If, for example, the southern and northern subduction zones bracketing Puerto Rico and Hispaniola are each active for 400–500 years, then hazard analysis based on a 500-year historic record could be grossly misleading. As is often the case, scientists cannot hope to assess the hazard associated with future earthquakes until they develop an understanding of the fundamental physical and geological processes in play.

The largest known earthquake in the entire Caribbean occurred just offshore of the north coast of Hispaniola on August 4, 1946. It was preceded by three years by another substantial temblor: an M7.5 earthquake offshore of northern Puerto Rico, in the Mona Canyon, on July 29, 1943. The 1946 event was larger, with a magnitude of approximately 8.0. This earthquake is thought to have been a subduction zone event, rupturing some 200 kilometers along the slab of Atlantic Ocean crust that is sinking beneath Hispaniola. A tsunami struck the coastline immediately following the shock; nearly 100 people drowned in the village of Julia Molina (now Mantanzas). The tsunami height was relatively modest—about 2.5 meters—but the shore region is flat and the wave pushed unusually far inland. According to some reports, as many as 1,800 people lost their lives as a result of this tsunami. Smaller tsunami were reported for days after the main shock; some of these may have been caused not by aftershocks per se but by underwater landslides triggered by aftershocks.

The 1946 quake emphasizes an important point about subduction zone earthquakes in the Caribbean: the offshore location of the subduction zone will blunt the shaking from large earthquakes occurring there, but subduction zone earthquakes are, as a rule, associated with tsunami. And as has been illustrated many times in the Caribbean and elsewhere, tsunami can be deadly even when the usual sorts of earthquake waves are not.

Less than two years after the 1946 earthquake another large earthquake, estimated to be magnitude 7.3, occurred along the coast of eastern Hispaniola. A last significant earthquake in this 20th-century cluster occurred on May 31, 1953, along the coast of northern Hispaniola. Once again, detailed analysis of earthquake data suggested a subduction zone event, albeit more modest in this case: magnitude 7.0. Having sprung to life with a vengeance in the middle of the 20th century, the subduction zone offshore of northern Hispaniola settled down fitfully through the rest of the century, producing seven moderate earthquakes, none of them rivaling the large events of 1943–1953.

In many ways, earthquake hazard on Hispaniola parallels that of other islands. The large islands of the Caribbean are bounded by major offshore plate boundary faults that represent a substantial hazard by virtue of their capacity to generate not only strong shaking but also severe tsunami. The beautiful, balmy Caribbean coasts are, moreover, far more extensively developed, and more densely populated, than they were when earlier large tsunami struck. Hispaniola faces these risks and more: here the plate boundary system comprises major fault strands that run on-land, through the island itself.

Efforts to investigate faults in the Caribbean—and to understand the hazard they pose—have lagged similar work in the United States (the western United States in particular). But unlike central and eastern North America, the populated part of the Caribbean is entirely an active plate boundary system. As such it poses a scientific challenge because, acre for acre, the modest Caribbean plate has more than its share of complexity. It also clearly poses a substantial practical challenge, and impetus, by virtue of the large and growing population at risk from future large earthquakes and tsunami in the region. One small silver lining of the 2004 Sumatra disaster is the heightened awareness it generated of tsunami risk along other subduction zones. With supplemental funding to augment ongoing U.S. Geological Survey efforts, tsunami warning systems in the Caribbean were slated for upgrade at the time of this book's publication.

10

Tsunami!

The earth is always shifting, the light is always changing, the sea
does not cease to grind down rock.
> —James Baldwin, in *The Price of the Ticket: Collected
> Nonfiction, 1948–1985* (New York: St. Martin's, 1985), 393

The 1867 tsunami described in the previous chapter was, as the world has recently witnessed, scarcely an unusual event. Nor was the scene of destruction
that followed.

Elsewhere in this book we emphasize how the world's rush, since the 1950s,
to expand the size of cities has been driven by an increase in global population. Like a box with flexible sides, the city expands to embrace all those
who favor the convenience, bustle, and economic opportunities of urban
life. When lateral expansion is no longer feasible, as in the walled holy city of
Bhaktipur in Nepal, or the confined economic and cultural island power-
house of Manhattan, the city expands upward. When both lateral and upward expansion are confined, the size of dwelling units inevitably contracts.
Few citizens leave these urban black holes, and when they do, they invariably
choose to swell the ranks of another city.

Yet one other type of place on our planet has beckoned since ancient
times—coastlines of continents, especially the earth's temperate and tropical shores. It has been estimated that 400 million people live within 20 me-

ters of sea level and within 20 kilometers of a coast, many of them within a few kilometers of the beach. Precise numbers are difficult to pin down because census compilations rarely list a household's height above sea level or its distance from the sea. Some idea of mankind's curious predilection to gravitate shoreward can be obtained by viewing the earth from space on a moonless night. Seen from above, the coastlines of continents and islands are illuminated festively by electric light bulbs (figure 10.1).

The attraction here is not so much the views nor even the fish: coastlines are trade routes and, being the termini for the world's rivers, streams, and subsurface aquifers, are nearly always endowed with a bountiful supply of freshwater for agriculture, as well as for thirsty populations and industries. This, of course, is why many of the world's largest cities are seaports: London, New York, Karachi, Calcutta, Hong Kong.

Left to themselves, the coastlines between cities evolve as a strip of urban development. In the industrial nations, where shoreline views command a premium price, strict regulation has protected some coastlines as wildlife refuges and recreational areas. In the developing nations, despite best intentions, shantytowns have sprung up along many beaches, and little by little invade remaining spaces between coastal roads and the shore. Temporary shelters grow by accretion into semipermanent structures, which eventually, through sheer weight of numbers, establish a right to permanent, if precarious, existence.

The coastlines of the world thus constitute a very special kind of city, a two-dimensional urban agglomeration, a mix of the world's poorest and wealthiest citizenry, destined by geography to occupy a thin band encircling a hinterland of landlubbers. The two-dimensional nature of this vast shoreline community is dictated by the laws of physics. The gravitational attraction of the earth holds the ocean firmly in its basin, with a certainty that humanity accepts without much thought. The gentle pull of the sun and moon causes the oceans to rise and fall like clockwork. Sea level has risen significantly since the end of the last ice age, but at a languid pace: the massive 120 meters or so of sea level rise in the past 8,000 years has flooded ancient shores at a no more than a fraction of a millimeter each year, enabling coastlines to keep pace with its advance: sand dunes and sandbars retreat, mangroves migrate, and deltas adapt. Every few hundred years, early civilizations packed their bags and rebuilt on higher ground with no more vigorous protest than, occasionally, a host of biblical legends and other oral cultural traditions about a global flood.

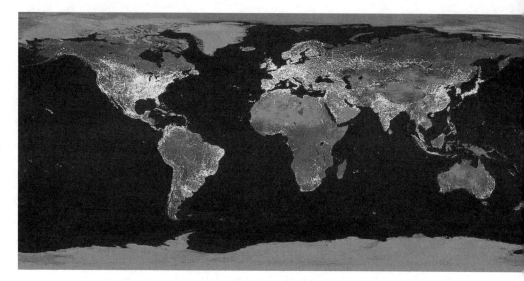

FIGURE 10.1. This composite image of a (cloudless) earth from space reveals the illumination of the world's cities—and coastlines in particular—by electric lights. (Data and image courtesy of NASA and National Oceanic and Atmospheric Administration)

Coastal dwellers grow accustomed to the occasional tempest, and the sea would lose much of its romantic attraction were it regularly as sedate as a millpond. No wind means no surf. Vexing winter storms and the rare inbound cyclone cause the sea to rise and surge inland, yet when it does so, it is usually preceded by unmistakable and adequate warning. The sky darkens, fishing boats head for harbor, and shore dwellers board over their windows or take cover on local hills or, in flood-ridden, hill-starved Bangladesh, on cyclone platforms. Ocean storms and surges are now predictable from satellite images, making possible timely coastal evacuation. Even the worst storm has a beginning and an end—stories with drama but few real surprises.

A tsunami, however, is a disaster whose arrival can come without warning, and whose effects are largely uncertain. Tsunami, meaning "tidal wave" in Japanese, are not the familiar tide-generated waves, but rather are generated by a sudden change in the level of the seafloor. They are usually caused by an earthquake or by the collapse of a mountain-sized pile of loose sediments deep beneath the ocean. Many cubic kilometers of seawater are heaved from their accustomed equilibrium, creating gigantic waves as gravity attempts to pull the sea back to its former level. In some cases the sea, eager to occupy newly lowered land, is heaved onshore, flooding the newly depressed shoreline. In other cases the sea is heaved oceanward, leaving fish flopping on the muddy, drained foreshore, only to be ripped up and thrown on land tens of minutes later when the waters overshoot and rush angrily back onshore.

Nowhere is the warning of a tsunami shorter than near the origin of an earthquake, especially if the earthquake has occurred at a subduction zone coastline, at the interface between a descending oceanic plate and the over-riding land. Here the sea either recedes or surges almost instantly onshore, flooding newly lowered land area. The on-land shaking during the quake destroys or damages houses, and survivors struggling to extract themselves from ruins in the succeeding tens of minutes can meet an inbound wall of water from the tsunami. The power of this fast-moving water lifts pebbles, trees, boulders, freighters, cars, warehouses, and walls with ferocious indifference, grinding and driving them inland until the wave loses its energy. It is this maelstrom of detritus that makes human survival of a large tsunami improbable.

Tsunami in History

Ever since the formation of the world's oceans, tsunami have repeatedly sloshed water over the edges of the continents. Meteorites, earthquakes, volcanic explosions, and the collapse of the unstable edges of continents and islands—in short, anything that makes a big splash—will generate a tsunami. Tsunami register in human history only when they are big enough to cause damage, yet small enough to leave survivors to describe them.

The volcanic explosion of Thera, a caldera now edged by the Mediterranean island of Santorini, 3,600 years ago was so catastrophic that it remained a legend until archaeologists and geologists exhumed the remnants of the Minoan cities and ports that were damaged by it. An inferred massive tsunami is thought to have been generated by the explosive phase of this eruption, which would have been sufficient to damage fleets anchored throughout most of the eastern Mediterranean. However, precise dates of ash deposits from the eruption indicate that the Minoans pottered on for a further century before being replaced by Mycenaean rulers. One possible reason for the brief continuing survival of Minoans on nearby Crete is that the island was not in the direct path of the Thera tsunami. The loss of life in the Thera eruption will never be known, but some idea of the numbers can be obtained from more recent examples. In 1883, the volcano Krakatoa exploded with about equal ferocity, creating a 30-meter wall of water that caused more than 36,000 deaths by the time it had traveled across the world's oceans.

We have already related the catastrophic multiple effects of the 1755 Lisbon earthquake and ensuing tsunami (chapter 3) as well as of the 1867 tsunami that struck the Caribbean. Tsunami are regular unwelcome visitors to the coasts of Japan, where in the past 1,300 years they have resulted in an estimated cumulative death toll exceeding 150,000, including five tsunami that have each claimed more than 10,000 lives. But nowhere in the world do we have account of a tsunami causing such damage and loss of life as the one that occurred in Southeast Asia on December 26, 2004.

Sumatra—Sunday Morning, December 26, 07:59

Great earthquakes have occurred repeatedly along Sumatra's coastline in the past two centuries. The 1833, 1861, and 1907 earthquakes resulted in tsunami whose antiquity had lulled local historians into a state of forgetfulness, and were the stuff of legend to most of those who lived within striking distance of the M9.0 earthquake that occurred at 07:59 local time on December 26, 2004. (Detailed studies suggest an even larger magnitude: 9.3.)

The first that the local population of Banda Aceh knew of the earthquake was a series of tremendous jolts and crashes that knocked many off their feet one minute before the 8 o'clock morning news. For many these were the last sounds they would ever hear, since houses over a large part of the region were demolished by the massive vibrations. Others, in stronger houses, in streets, motor vehicles, and boats, barely had time to realize, above the shouts and screams and groaning of buildings, that an earthquake was under way when they heard a sound that many describe as a roar like a jet plane taking off. Those near the beach were soon engulfed by the cause of the sound: an incoming wall of water more than 3 meters high, and in some places as high as 25 meters. For reference, in places, this wall of water was one quarter as tall as a football field is long. Those as far as 9 kilometers inland, viewing from upper levels of reinforced concrete structures, watched in terror as a black, oily-looking, turbulent stream pushed and rolled trees and soil, sand and mud, cars and boats, animals and people inland, northward and eastward from the epicenter.

The inward-bound surge continued near the epicenter for ten seemingly endless minutes with terrible consequences and vast reach. As captured by a few observers wielding video cameras, the thick, black water poured inland with no apparent intention of slowing. It flooded fields, villages, and roads, carrying almost everything in its path, armed to the teeth with suspended debris, and completely indifferent to trees and structures. Buildings shaken by the earthquake were scoured off their foundations and added to the moving mass. Few people caught in the surge survived unless they were able to climb to the safety of trees or intact structures above the water level. Then, ponderously and more slowly than it came, the water receded from the higher parts of the mainland, leaving the heavier debris in scattered heaps of widespread devastation that few humans have ever witnessed, even in times of war: broken bricks, concrete fragments, mangled cars, battered boats, and the detritus of a hundred thousand households. As the water ebbed, it carried lighter effluent—paper, sewage, fish, and people, dead and alive—sucking them back to the ocean or dropping them in pools and piles on the way.

The earthquake unleashed powerful forces that had previously supported the land at its previous level. Parts of the former Sumatra shoreline were now, only 20 minutes after the earthquake, permanently below sea level, with the foundations of former marina structures and sea-view hotels immersed seaward of the new coastline. The terror did not end there: those fortunate enough to survive the earthquake and tsunami unscathed were to face many days of searching for relatives amid the stench of decaying corpses, a seemingly relentless barrage of M6 aftershocks, and the constant threat of further tsunami. Most survivors had lost one or more relatives. In many cases whole families were lost.

Nicobar Islands, 400 Kilometers North, 08:03

The main rupture zone of the M9 earthquake unzipped 650 kilometers of the north-northeast-trending Indian plate boundary in roughly four minutes. Then the rupture slowed, but did not stop. Over the next six to seven minutes it ruptured a further 650 kilometers of the plate boundary, continuing on to islands north of the Andamans and creating one of the longest single earthquake ruptures in recorded history (1,600 kilometers). Curiously, the seismic waves that rattled Port Blair and signified the initial rupture of the plate boundary had passed quickly northward before the bulk of the Indian plate slipped eastward and downward slowly beneath the Andaman Sea. As a result the harbor at Port Blair began to sink, and the town slowly flooded, several minutes after the shaking ceased and tens of minutes before the arrival of the tsunami from the south. The halfway point where rupture slowed, or hesitated, was a patch of the plate boundary near Car Nicobar Island that had itself produced a tsunami following an M7.9 earthquake almost exactly 125 years earlier, on December 31, 1881.

Eyewitness reports of shaking on the western Nicobar Islands are rare because so many people perished. The islanders on Katchal, near the rupture zone but more than 400 kilometers north of Banda Aceh, were shaken violently a few minutes past 8 o'clock and, while recovering from the shock, must have seen the 10–15-meter-high tsunami surging toward them. Unlike the people of the mainland, however, many of those on Katchal lived in a crescent of villages on a terrace of coral near sea level , surfaced by soils and palm trees. Most of the pre-earthquake Katchal population of more than 5,000 were swept away. A few days later a mere 100 survivors were evacuated. Only a single person was confirmed dead; the sea had carried off almost the entire population, and with them many of their bamboo and thatch homes.

Islanders on the east-facing parts of the Nicobar Islands fared better than those on the west-facing parts, because Katchal Island took the brunt of the northeast-directed tsunami, but none of the islands escaped the wave. The Nicobar Islands, moreover, all sank—the same fate as western Banda Aceh. Sand spits and sandbars no longer joined offshore coral clumps to the mainland. Trinkat Island was cut in two by the submergence of the narrow isthmus that once linked its northern and southern halves. Estuaries of small, nonperennial streams now became inlets of the sea. Shallow harbors became deepened. The area of low-lying islands lacking sea cliffs shrank by 10 percent, and the larger islands became refuge centers for the population evacuated from the smaller islands. The southernmost point of India, Indira Point (formerly Pygmalion Point), was flooded, and vegetation was ripped and removed by the tsunami, leaving the damaged 35-meter-high cast iron lighthouse surrounded by waters of the Indian Ocean.

A tragedy in the smaller islands was not only the wholesale loss of coconut palms, a source of food and drink, but also a major reduction, both temporary and permanent, of underground supplies of freshwater. Drinking water on low-lying islands exists naturally only in the form of a fresh-water lens maintained like a liquid iceberg, an eighth of it above sea level and seven eighths of it below sea level, in the body of sand grains and rock fissures inland. The lens is thickest in the center of the island and tapers to nothing near the edges, where freshwater seeps into the sea. Islanders pump water in moderation from wells that pierce the lens inland—not too deep, because beneath it lies saltwater, and not too much, because the lens, once gone, takes many years to replenish. A rapid rise in sea level caused by sinking of an island expels a corresponding fraction of this freshwater, and the salty inundation from the tsunami contaminates the remainder. It will take many years for rainwater to flush away the salt.

Car Nicobar, 600 Kilometers North, 8:06

In 1881, following an M7.9 earthquake, the low-lying island of Car Nicobar was raised in the west and tilted slightly down to the east just before the arrival of a small tsunami. A similar tilt occurred in 2004, but the shaking from the earthquake and the amplitude of the 2004 tsunami were much larger, causing the military bases there to urgently telephone headquarters on the mainland that coastal tracts on the island had been inundated and harbor facilities destroyed. The message seems not to have been received, or not to have been believed, by those in a position to realize its import, since this surely would have resulted in a warning being issued to mainland India and the coasts of Thailand toward which the tsunami was now speeding. Some reports claim that the military bases on the island lost all communications with the mainland, an unlikely scenario for a military command purporting to be located in the islands to defend India's strategic interests.

The Oceanic Tsunami

As the tsunami traveled away from the shore, it encountered rapidly deepening ocean water. Since the speed of a tsunami wave is proportional to the square root of the depth of the ocean it traverses, it doubled in velocity for every quadrupling in depth, hitting a maximum of around 800 kilometers per hour some 100 kilometers offshore. Accompanying its increase of velocity was an increase in wavelength to hundreds of kilometers, and a decrease in amplitude to tens of centimeters. For this reason ships would have noticed the earthquake (ship captains described a grinding noise on the hull of the ship), but not the tsunami, whose midocean amplitude was far lower than the height of wind-driven waves, and hundreds of times smaller than the surges about to hit the surrounding coastlines.

Serendipitously, however, a number of oceanographic satellites crossed the sky in the hours following the earthquake and were able to track the killer tsunami crossing the ocean within tens of minutes after the wave hit the mainland (figure 10.2). Future descendants of these remarkable satellites may well provide the method of choice for global tracking of future storms and tsunami.

The tsunami focused most of its energy as a broadside wave traveling westward and eastward, with lesser energy focused to the northwest and southeast. As the wave approached coastlines, the energy of the midocean tsunami was concentrated into greatly increased amplitudes and shorter wavelengths. In Thailand, what had been an invisible, large-scale swell in midocean became a wall of breaking waves with amplitudes in places exceeding 10 meters. A yacht 1.6 kilometers offshore recorded the passage of the tsunami on its depth sounder as it rose and fell 6 meters in 12 meters of water. (This wave, although large, was still spread out over a large wavelength and thus would have merely lifted the yacht up and set it back down gently.) Nearby, a couple of scuba divers near the ocean floor felt the swell of the tsunami as it passed, but were unaware of the disaster unfolding on land. Photographs show a wall of water advancing rapidly toward the Phuket shore with terrified people in the foreground.

Distant Shores

It is tragic to realize that of the many thousands of people who died in the tsunami, some at least may have heard the news of the earthquake on the radio more than an hour before the deadly tsunami arrived at their local shores. Some may even have looked their approaching death squarely in the face as they viewed Web pages linked to the U.S. Geological Survey global earthquake display, which gave the earthquake's location and magnitude, and which, if they had a modest amount of education, would have warned them to head for the hills and not for the beach that day. Although scientists attempted to alert local authorities to the possible arrival of a tsunami, the absence of detailed data and a clearly formulated action plan made it impossible for officials to take such a warning seriously. Thus the wave arrived without warning throughout all the distant shores of the Indian Ocean: Malaysia, Thailand,

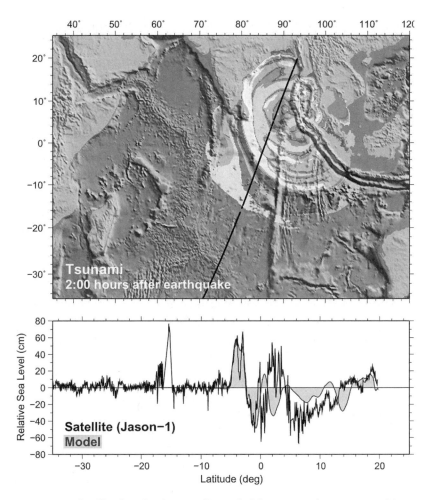

FIGURE 10.2. Satellite-based estimates of water height were used to construct this map of the Indian Ocean tsunami two hours after the earthquake. By this time the leading edge of the wave had traveled across the Indian Ocean, causing widespread destruction in Sri Lanka and southern India. (Courtesy of National Oceanic and Atmospheric Administration)

Myanmar, India, Sri Lanka, and then on to the Maldive Islands, the volcano Reunion, and the Seychelles, with a final dose of death to the shores of Somalia, Tanzania, and Kenya on the east coast of Africa.

After its fatal rampage the tsunami did not stop, but reflected from and refracted around coastlines, traveling at jet speeds but with negligible power to destroy, past the southern tip of Africa and up the Atlantic to finally rock the icy Arctic Ocean with waves less than one inch high, there meeting with waves that had traveled the eastern route past Australia and up the Pacific Ocean through the Bering Strait. The Americas felt the spent dregs of the tsunami on both their eastern and their western shores, with occasional focusing on favorable coastlines, as in Mexico, where quirk waves were recorded with almost 1-meter amplitude. When all the waves had subsided, sea level had risen globally a fraction of a millimeter because of seawater displaced from the shallowed depth of the Bay of Bengal.

The 2004 tsunami could scarcely have been worse. The fault moved in such a way that a massive wave was created. Almost all great subduction zone earthquakes will generate a tsunami, but not all tsunami are created equal. When a massive M8.7 aftershock—a great earthquake in its own right—struck on March 28, 2005, to the south of the 2004 quake, it generated a markedly modest tsunami (see sidebar 10.1). In this case, shaking during the earthquake caused most of the death and destruction.

The 2004 tsunami, however, was not just enormous: it also could not have come at a worse time. Had it been at night, many of the children would have been spared, since surely they would have been inland, tucked in bed. Had the tsunami occurred at low tide, as it did in parts of India, its fury would have been at least somewhat subdued. But for most of the east coast of India, the tsunami occurred at high tide, aggravating its flood potential. Had the tsunami occurred on a schoolday, fewer of the local residents and none of the children would have been on the beach. But it occurred not just on a weekend, but on a festival weekend, the day after Christmas. More touristy beaches throughout the region were crowded with visitors from around the world. The tsunami could hardly have been larger, because the earthquake involved

SIDEBAR 10.1

In the aftermath of the M8.7 quake of March 28, 2005, some news media articles described a "debate" among scientists. Was the quake an aftershock or was it a great earthquake? In fact, it was both. By all accounts the earthquake fit the criteria that define an aftershock: it was close to the main shock in time and space and was clearly an event "set into motion" by the earlier large main shock. The point is that aftershocks *are* earthquakes, and sometimes they are big earthquakes. Every once in a while, they are even great earthquakes.

slip of the entire plate boundary from Sumatra to the Andamans. Historical precedents along this part of the plate boundary were for earthquakes smaller than magnitude 8. No scientist would have dared suggest an M9.0 earthquake, or that its possible effects might be combined with every other imagined worst-case scenario. It was, and for that matter still is, simply too unlikely. But sometimes unlikely events happen.

The Cost

The precise death toll from the 2004 earthquake and tsunami will never be known, but it approaches 300,000—including those who died from diseases and other causes in the aftermath of the immediate tragedy. Indonesia alone lost 115,000 people, Sri Lanka more than 29,000, India more than 15,000, and Thailand 5,000. Hundreds of tourists lost their lives as well. A statistic that is small in the greater scheme of things but somewhat astonishing to consider: the 2004 Sumatra earthquake and tsunami caused the single largest loss of life among Swedish citizens in any natural disaster in recorded history.

The cost of reconstructing the source region in Banda Aceh has been estimated at $4 billion. The cost of fixing the roads, and dwellings, and communication lines of all the Indian Ocean shorelines is almost certainly much more. But from almost the first moment that news of the devastation reached people around the globe, there seemed to be no question that funds would be allocated, donated, and borrowed to reconstruct the damaged areas.

Hope Amid the Ruins

It is difficult to look death and destruction squarely in the eye and return to the optimistic theme of rebound. Indeed, one could scarcely point to a better example of the staggering human toll that earthquakes and other natural disasters sometimes take, society's remarkable resilience notwithstanding. Nor could one point to a better example of the tragedy of preventable disaster. Tsunami warning systems are not the stuff of science fiction, nor do they come at a prohibitively high cost. An effective early warning system for the Indian Ocean region would require only a modest expenditure of funds, perhaps $20–30 million, because much of the infrastructure for such a system is already in place. One of the most expensive elements of such a system has existed for decades: a worldwide seismic monitoring system. A warning system could also utilize existing radio communications systems, public warning sirens, and, perhaps most of all, public education through schools and the media. Even without a more sophisticated, high-tech warning system developed with substantial new instruments, a cost-effective, low-tech system would be tremendously effective.

No amount of future investment can, of course, heal many of the wounds from the 2004 disaster. A tragedy of this enormity will require decades of recovery efforts preceded and followed by changes in the way that local authorities view construction in hazardous regions.

And yet, the recovery effort continues, fueled by the usual infusion of global resources. Affluent nations and citizens of the world pledged $5 billion in emergency assistance in the immediate aftermath of the disaster. Tourists from 24 countries were killed on the beaches of Southeast Asia, but within the first weeks of the earthquake, aid had been contributed by 51 nations. Aircraft brought metric tons of water and produce in the first weeks following the earthquake and tsunami. Survivors were concentrated in camps, tourists were repatriated, and local wars were (but in some cases only temporarily) forgotten. Nations spoke of canceling the foreign debts of many of the afflicted nations, since payments on national debts cost developing nations millions of dollars each day.

Viewing beaches in the tsunami-afflicted regions after the first week became a spectator sport for the curious and lucky hinterland dwellers. Journalists, dignitaries, scientists, and others rushed to the area, in part to do their jobs but also in part to *be there*: to witness firsthand the power of forces that defy human imagination. Individuals in the tourism business expressed dismay over the loss of revenue and jobs in the immediate aftermath of the disaster, yet one can say with near certainty that the tourists will return. Local laborers, both skilled and unskilled, will be pressed into service in the rebuilding effort well before the tourists return. Their talents will be put to use around the rim of the Indian Ocean, reconstructing dwellings and infrastructure.

Earthquake disasters play out of timescales of seconds to minutes; tsunami disasters unfold more slowly, over timescales of hours. Elastic rebound as manifested within human societies is a longer-term process. The most heavily impacted regions in Southeast Asia will not bounce back right away, nor will they bounce back without more suffering than was experienced in the initial weeks following the disaster (figure 10.3). The rim of the Indian

FIGURE 10.3. Devastation in Banda Aceh in the wake of the 2004 tsunami. This location is over 3 kilometers inland from the shore. (Photograph by José Borrero; used with permission)

Ocean has been changed forever by the events of December 26, 2004. Coastlines and other landforms will never again be as they were; nor, one imagines, will the lives of the survivors ever be the same.

It is all too easy to imagine that no good can come from such a horrific and far-reaching disaster, but even this great earthquake can offer a silver lining to its survivors. Afflicted cities will be rebuilt safer than before, public warning systems and response plans will be posted along numerous shorelines, and earthquake-resistant construction will be taken more seriously in many parts of the world. The most important long-term effect of the 2004 Sumatra disaster might just prove to be education. The lesson exacted a staggering price from those directly in the path of the quake and its tsunami, but it has generated a more effective earthquake- and tsunami-education program worldwide than any effort that could ever have been planned.

11

City of Angels or Edge City?

... a delightful place among the trees on the river. There are all
the requisites for a large settlement.

> —Fr. Juan Crespi, August 2, 1769, in *A Description of*
> *Distant Roads: Original Journals of the First Expedition*
> *into California, 1769–1770*, translated by Alan K.
> Brown (San Diego: San Diego State University
> Press, 2001)

Although this book focuses on societal response to earthquake disasters, many common threads can be found in societal response to other types of disasters. Some regions seem especially prone to disasters of all shapes and sizes, perhaps none more so than southern California, which can be star-studded and star-crossed in equal measure. This chapter steps away from the specific responses of societies to one type of disaster to instead consider the response of one society to myriad disasters.

In southern California, disasters sometimes seem to pile up like, well, cars on a southern California freeway. During one memorably miserable week in October 2003, for example, firestorms laid waste to almost 700,000 acres in the region—2,000 homes, 24 lives, and a staggering $2 billion in property damage.

It was a little like an earthquake in slow motion. The 1989 Loma Prieta earthquake had claimed about three times more lives (63) and total property damage ($6 billion), but the number of homes rendered uninhabitable by that powerful temblor was lower (1,450) than the number destroyed by the firestorms of 2003. That the disaster played out slowly, over the span of several days rather than several tens of seconds, was a curse as well as a blessing. Advance warning kept the death toll from climbing higher; it also generated high anxiety among tens of thousands who would not lose their homes as well as the few thousand who would. Fires are less kind than earthquakes in another critical respect as well: they can reduce an entire house and its contents to ash, whereas much can often be salvaged from even a severely earthquake-ravaged home. Fires can even have their own aftershocks, after a fashion: heavy Christmas Day rains turned parts of two burn areas into torrents of fast-moving debris that swept through two campgrounds and claimed 16 lives, most of them children. Even heavier rains in early 2005 caused a more massive landslide in the coastal community of La Conchita.

Before the flames were even extinguished in 2003, stories appeared in the media describing the conditions that made such disasters almost inevitable. Most fundamentally, population pressures inspire an ever-growing push to inhabit land at the ragged edge of hospitality to human habitation. Even without mischief supplied by human folly or malice, forest fires are to forests what disease is to biological populations. Indeed, fires are such an integral element that some long-term forest cycles are predicated on the occurrence of occasional catastrophic conflagrations. The seeds of certain plants will not germinate unless they are first charred and then watered by subsequent rains.

As discussed at length by Mike Davis in his book *The Ecology of Fear*,[1] 20th-century Californians approached fire danger with the same brashness that they brought to other challenges that threatened to disrupt their intent to create paradise on earth. Forests, forest fires—clearly such things could be *managed* in a way to allow for peaceful coexistence between people and natural hazards.

It didn't turn out that way. By now it is old news that seemingly sensible efforts to manage the forests went very badly awry. Without human intervention, forests experience lots of small fires. Paradoxically, these small disasters render large disasters less likely by creating natural firebreaks—places where the next fire will likely run out of fuel. Here again one finds some degree of similarity with earthquakes: if one plots the number of earthquakes of different sizes in a given region, one finds a curve that is virtually identical to a plot for fires (figure 11.1). This starts to sound like an explanation of what is in fact a common misconception: that small earthquakes help prevent big earthquakes. But one can't take the fire–earthquake analogy too far. Plate tectonics forces provide a certain "energy budget" for earthquakes: as the plates move, energy is stored, or banked, in the crust. This energy reserve must, over the long term, be spent. Spending the entire earthquake budget on small earthquakes would be rather like depleting lottery winnings through buying candy from gumball machines. If the earthquake budget is a lottery jackpot, small earthquakes are nickels, and the budget is simply too big to be spent a nickel at a time. As far as the bank balance is concerned, nickels don't count.

But small fires do matter. Without human intervention, large forest fires will erupt in a given area at a given average rate. With well-meaning but misguided humans around to squelch the small fires, large fires become more likely. The best of intentions have gone awry in other respects as well, where forest fires are concerned. As expressed by California Department of Forestry and Fire Protection District Chief Tom O'Keefe, at the dawn of the 21st century, southern Californians had "saved the trees but lost the forest."[2] By this he meant not only the policy of fighting small fires, but also other long-term conservation efforts that failed to allow trees to die a natural death. Here one can appeal to the analogy of animal species: if any population multiplies beyond the limits of its resources, the stage is set for catastrophic decimation by communicable disease.

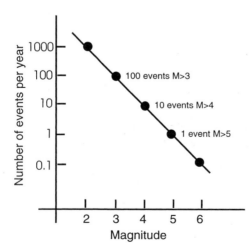

FIGURE 11.1. A b-value curve indicates the number of earthquakes of each magnitude or greater in a given time. While the precise numerical values differ considerably in more and less active areas, the straight line—with a b-value close to 1.0—is observed almost universally.

In southern California, fire season arrives almost as predictably as does hurricane season in the Atlantic Ocean. The fire cycle typically begins in the winter—the first three calendar months of a year in particular—during which time the lion's share of annual precipitation commonly falls. As the song says, it never rains—for nine long months each year it can literally not rain a drop in southern California. And then for three months it pours. Not constantly, but in fits and starts—fits and starts of sometimes nearly biblical proportions. Individual rainstorms soak the area with as much as five inches of rain in 24 hours; with an average annual rainfall of about 15 inches, two such storms can make the difference between an average year and drought conditions.

The official rainfall statistics for Los Angeles reflect the amount of precipitation measured at City Hall in downtown LA. As one would expect, it usually rains much harder and longer along the windward side of the San Gabriel Mountains. At an elevation of about 5,700 feet, Mt. Wilson receives an average of over 40 inches of rain each year, 80 percent of it between December and March. Were this seasonal rate to continue throughout the year, the annual rainfall on Mt. Wilson would comfortably exceed the amount required to define a region as a rain forest.

As winter rains soak into the ground, parched valleys and mountains respond with tremendous enthusiasm. By April, the dry, golden California hills are transformed: trees shine as if freshly polished, grasses and weeds grow like . . . well . . . weeds, and blossoms appear on plants that didn't seem to have enough imagination to think about blooming.

Then the rains stop. Through April, May, and June, a heavy marine layer typically keeps much of southern California under a blanket of moisture and cool temperatures. Although it may not be getting watered, the ebullient early spring flora does not suffer too badly through the days of June gloom. Suffering comes later, when baking temperatures arrive in July and August. The last of the blossoms wilt and drop; green turns quickly to burnished gold.

Then it gets hotter. In September and October, massive high-pressure systems commonly develop over the western United States. These systems block the usual onshore flow of moist, buffered air from the sea. Worse yet, they stoke winds in the other direction, winds that are known formally as Santa Anas and informally as devil winds. With wind speeds from 30 miles per hour to as much as 60 miles per hour, the origins of the informal name are not difficult to understand. (The association with sainthood is another matter.) These winds kick up high surf, damage crops, and topple trees; at their worst they pose a hazard to large vehicles such as big rigs and campers. They also fan flames, with a vengeance.

Santa Ana winds do not, as a rule, start fires. The weather condition is hot and tinder dry: no rain, no storms, no lightning. But the absence of atmospheric electrical disturbance is small consolation: the devil winds cannot start fires, but they surely can fan the flames. Flames, in turn, are all too easy to start; in an urban setting, accidents and arson often provide the spark. And among those with a pathological bent, fire sparks fire; perhaps the most disheartening part of the equation is the rise of "copy-cat" arson incidents once large fires are already burning.

The Angeles National Forest and the San Bernardino National Forest each comprise as much acreage as the state of Rhode Island—and these two enormous parcels do not account for all of the mountainous wilderness in southern California—not by a long shot. Over this vast extent of real estate the winter rains fuel explosive growth of flora that the summer sun bakes dry. When the devil winds of autumn bring even hotter, drier, and windier conditions, the only missing piece is the spark—but in an area the size of southern California, this piece is all too easily provided by negligence or malice.

And so, with sickening inevitability, southern California burns. Catastrophic fires do not erupt every year there, but perhaps once a decade. Indeed, in late October 2003, almost exactly a decade had passed since the last major firestorm in southern California. The fires of 1993 consumed less total acreage than those of 2003, but the biggest fires burned closer to central population centers, in the mountains directly above Altadena as well as in Malibu and Laguna.

The early 1990s were not kind to southern California. The M7.3 Landers earthquake of June 1992 focused its destructive energy in a sparsely populated part of the Mojave Desert, but waves from this powerful temblor were a strong and rude awakening minutes before 5 o'clock on that summer Sunday morning. Half a year later, El Niño-driven storms pummeled Los Angeles, pelting the area with almost 26 inches of rain over four months and wreaking havoc with property and hillsides alike.

The next disaster was anything but natural: a week of violent rioting following the acquittal of four white police officers whose show of force against black suspect Rodney King had been documented on videotape—and replayed about 6 million times on television sets around the planet. Finally, the African-American community said, hard evidence of excessive force on the part of law enforcement officers. Not guilty, the white Simi Valley jury proclaimed. It wasn't hard to understand the rage.

At the time of the Lisbon earthquake of 1755, many people were inclined to see great earthquakes as the handiwork of a powerful God. A quarter of a millennium later, people are far less inclined to see the hand of God behind what we recognize to be unpredictable natural events. It can thus only be ascribed to coincidence alone that the Northridge earthquake struck less than two years after the Rodney King verdict and within a dozen miles of Simi Valley—just after 4:30 in the morning on Martin Luther King Day. With a magnitude of 6.7, the Northridge earthquake was by no means a great earthquake, but much of the strongest shaking from this powerful temblor was focused in the densely populated San Fernando Valley and northern Los Angeles basin. When the dust settled, Angelenos began to tally the toll: 57 deaths, 4,000 houses red-tagged, tens of billions of dollars in property losses. As of 2003, the Northridge earthquake still stands as one of the most expensive natural disasters in U.S. history, second only to Hurricane Andrew.

"They don't call it Edge City for nothing."[3] In the aftermath of the Northridge earthquake, this was the observation of an eminent seismologist who had chosen not to live in southern California. Coined by *Washington Post* journalist and author Joel Garreau, the term "edge city" was not reserved for Los Angeles, but for any city that met five criteria, including a population that increases every workday morning and decreases every afternoon (indicating more jobs than homes). In fact, of the 123 edge cities Garreau identifies, nearly 20 percent are within the greater Los Angeles region. Along with the metro District of Columbia and New York City regions, Los Angeles is perhaps less of an edge city as strictly defined than a giant amalgamation of edge cities. An edge megacity, if you will.

Still, perched on the geographical as well as the metaphorical edge, the simpler moniker seems quite apt for Los Angeles: Edge City. There is indeed a compelling case to be made that, in building castles to the sky, southern California has dug its own grave. Homes and other structures encroach increasingly not only on a wilderness that cannot be tamed but also on landscapes that have never been stable.

A bluff near the ocean may offer a million-dollar view, but when a geologist considers a bluff, he or she thinks about prosaic details such as bedding planes. Bedding planes have nothing to do with sheets and blankets; the term describes the surface that separates two distinct layers of sedimentary rocks.

In a marine environment such as that found in the Los Angeles basin as recently as five million years ago, sedimentary layers usually deposit themselves in orderly fashion—which is to say, flat. Newer sediments progressively bury older sediments, resulting in the compaction of the latter and a progressive metamorphosis from loose goo to harder rock. If a thrust fault later develops under a layer cake of sediments, a flat cake can turn into an arched cake, with layers of sediments draped over a hill, the contours of the bedding planes mirroring those of the hill itself. The trouble is, bedding planes almost invariably represent relatively weak zones in between what has become harder rock. Especially if groundwater manages to lubricate the surface, bedding planes can become sliding planes, and entire hillsides succumb to the insistent force of gravity.

Worse yet, landslides represent only one of the serious gravity-induced perils in southern California. As alluded to earlier, the fire cycle turns out to be only part of a larger natural hazards cycle: after the mountains burn, their denuded top layers can turn to muck when the winter rains fall. The resulting downward-careening mess, known as debris flows, can carry enormous quantities of water, dirt, trees—even houses—down into foothill communities at high rates of speed. The rocky bits in a debris flow range in size from silty particles to large boulders. The addition of water and organic material, plus motion, creates a well-mixed slurry that can resemble wet cement, and that can travel at speeds as high as 40 miles per hour. In January 2005, a massive debris flow buried part of La Conchita, a bucolic (but especially unstable) part of the coast.

Debris flows are common in the foothills of mountains as rugged and steep as the San Gabriels. Recognition of this hazard has led local communities to build flood control basins and structures known as catch basins to trap material that would otherwise get washed down certain canyons and into foothill communities. These basins do their job—to a point. If a debris flow is large enough, flood control basins and catch basins can be overwhelmed. In 1980, a debris flow overran the stops and buried a community of San Bernardino homes in mud up to their roof lines. The winter storms of 1980 produced other serious flows, particularly below mountainsides that had burned the previous fall.

The images from 1980 were dramatic, but the last truly major debris flow in the Los Angeles area was in 1969. During the especially wet rainy season, debris flows were as common throughout the San Gabriel Mountains as spots on a leopard. Many of these were of only minor consequence, but below a prior burn area above the foothill community of Glendora, a debris flow unleashed mud, rocks, and muck to the tune of a million cubic meters. If that much muck is hard to imagine, picture it this way: enough for a 3-meter-deep blanket covering 60 football fields.

It is scarcely an exaggeration to say that in southern California, the most inviting building sites are the most perilous places to live. Living in the foothills means not only living with fire and landslide hazard, but also living atop an active fault—by definition, since the hills exist in southern California because active faults have pushed them upward. Life along California's scenic coastline involves other sorts of hazards, such as storms and landslides as well as, well, earthquakes.

The argument can be made—has been made, by authors such as Mike Davis—that California's precarious situation can only worsen over time. As population pressures lead to more development in the mountains, more homes will be at risk from fires, as well as landslides and debris flows. And then there is the conclusion that has become increasingly clear to earth scientists in recent decades: southern California has not yet been hit by The Big One. The Northridge earthquake was, if not small potatoes, only modest potatoes by seismological standards. As scientists have improved their estimates of the earthquake budget for the Los Angeles area alone, it has become increasingly clear that the budget cannot be spent on earthquakes smaller in magnitude than the Northridge quake. Both "budgetary considerations" and recent geologic investigations point to a grim conclusion: that earthquakes much larger than Northridge—perhaps as large as magnitude 7.5—have occurred in the past, and will occur in the future.

The greater Los Angeles region is crisscrossed by at least a dozen major faults, any one of which will not generate a major (magnitude greater than 7) earthquake more than every few thousand years (figure 11.2). The odds that any particular fault will produce The Big One over a given 100-year window are thus small. But to assess the hazard of the entire region, one must worry not about the odds of a big quake on a particular big fault, but rather about the odds of a big quake on any of the region's big faults. These odds are much higher; perhaps as high as one-in-three in a century.

What will The Big One be like? After decades of measuring and analyzing earthquake shaking, scientists do not believe that the highest-amplitude shaking will be much worse during a Big One than during a Pretty Big One such as Northridge. Although complex in detail, the reason for this is basically twofold. First, a very large earthquake will occur on a very large fault,

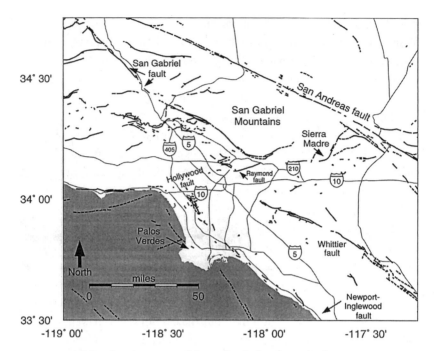

FIGURE 11.2. Map showing some of the major fault systems in the greater Los Angeles area.

but the entire fault does not move at the same time. Rather, a rupture pulse, or ripple, travels the length of a fault the same way (more or less) that a wrinkle can travel the length of a carpet. Second, at any given location near a big fault, the strongest shaking will generally come from the rupture at its closest approach to that location. Think of the earthquake as a big train. Standing still next to the track, an observer will hear the train as it approaches and as it leaves, but the loudest sound will be limited to the time when the train is passing directly by. The amplitude of the loudest sound will be about the same whether the train has 5 cars or 50 cars.

Like all of the analogies in this book, this one should not be taken too far. But it does capture certain key aspects of earthquakes and the shaking they generate. If the peak shaking does not get that much stronger during very big earthquakes (compared with moderately big earthquakes), some things are different—including the duration of shaking and the extent of the region exposed to severe shaking. An M7.5 earthquake might rupture 100 kilometers of a fault, such as the Sierra Madre fault at the base of the San Gabriel Mountains. In some ways this will be like six or seven Northridge earthquakes laid end to end. That is, instead of the strongest shaking being concentrated around a fault patch some 15 kilometers long, it will be strewn out along a much longer swath. Worse yet, at any one location, the total duration of potentially damaging shaking will be much longer than during a smaller earthquake. The performance of any structure during an earthquake depends critically on how long that structure is shaken. As seismic waves batter a structure, a couple of things happen, none of them good. Most notably, materials start to crack and yield, and buildings start to sway. As soon as the upper part of a structure begins to move sideways, it inevitably shifts away from the vertical support elements that were designed to hold the building up, generating potentially enormous stresses on these supports. Even if earthquake shaking is predominantly horizontal, the shifting mass of the upper building can pound the lower parts into submission.

Other types of failure occur as well. The lower floors of a high-rise building must be strong enough to support the upper floors, while the upper floors can be built with less strength. At some point in the structure, the strength of one level is lower than that of the level beneath it. This generates a point of weakness that can become a point of failure, causing a single floor to pancake in a building that otherwise appears almost unscathed. The Northridge earthquake struck in the wee hours of the morning, on a federal holiday, no less. People were, overwhelmingly, asleep in their beds at the time; most of them were in wood-framed structures that, both by design and by the intrinsic properties of the building materials, offer considerable resilience to shaking. People were also, overwhelmingly, with their families.

We have no way of knowing when The Big One will strike, not which year and not what time of day. The mind boggles at the thought of a Big One occurring at 8 o'clock in the morning on a weekday, at which time Los Angeles area freeways are enough of a disaster on any given day. If The Big One strikes at 2 o'clock in the afternoon, the freeways will still be busy, and many people will be working in buildings that are less safe than their homes. A lot of people will be anywhere from a few to a lot of miles from their loved ones, unable to reconnect when phone lines and freeways become equally choked and useless.

We also have no way of knowing on which fault The Big One will strike. Ultimately, as expressed many decades ago by Professor Nick Ambraseys, earthquakes don't kill people; buildings kill people. The extent of havoc wreaked by any one earthquake will depend critically on the inventory of structures around to be damaged—as well as the proximity of the earthquake to those buildings. In recent decades scientists have learned a lot about the giant geologic jigsaw puzzle that is southern California. Using methods akin to medical CAT scans and other techniques, we have not only identified the geometry of major faults but also have begun to investigate the prehistoric earthquakes that have occurred on them.

Some Los Angeles-area faults are scarier than others. An M7 temblor on the Raymond fault would tear a destructive swath through the San Gabriel Valley, generating its most destructive shaking in Arcadia, San Marino, and Pasadena. An M7.5 earthquake on the Sierra Madre fault would shake the daylights out of a larger part of the San Gabriel Valley. And so forth. One cannot point to any location in the greater Los Angeles area that could not experience extremely severe shaking from one fault or another. But no location is as densely populated as the central part of Los Angeles, and perhaps no fault is quite as scary as the Puente Hills blind thrust fault.

The Puente Hills fault is bad news for several reasons. First, there is the fault's central location: it extends from Whittier through downtown Los Angeles, and west into Beverly Hills. In describing the fault's location relative to the city of Los Angeles, University of Southern California geologist James Dolan likes to use the word "lurk."[4] Then there is the fault's geometry. As a blind thrust fault, it does not extend to the earth's surface; its upper edge is overlain by a thick blanket of sedimentary rock that is being warped by the long-term motion of the fault. This might sound like good news: when an earthquake strikes, the moving fault will be no closer than a few kilometers from the nearest inhabited real estate on the earth's surface. The problem is that a lot of inhabited real estate will be no farther than a few kilometers from the moving fault.

A major earthquake on the Puente Hills fault would focus its most destructive energy in proximity to the fault, in and around central Los Angeles. But the collateral damage could be severe as well. Consider the Northridge earthquake. Its strongest shaking was concentrated in and to the north of the San Fernando Valley, but even in the neighboring Los Angeles basin shaking was strong enough to collapse a section of Interstate 10 near Las Cienegas and to cause serious damage in Santa Monica, over 20 kilometers to the south.

The conclusion starts to appear inescapable: Edge City is doomed. It won't fall into the ocean; this misconception is easily dismissed with considerations of the tenets of plate tectonics and the geometry of the North American–Pacific plate boundary. Rather, Los Angeles is destined to fall apart and/or burn up, one Rhode Island-sized piece of real estate at a time. Any sensible person would realize this, and respond appropriately—as indeed they apparently do when modest disasters provide occasional wake-up calls. As Mike Davis pointed out, over half a million people packed their bags and left southern California in the few years following the Northridge earthquake. Property values, already battered by recessionary forces in the early 1990s, took a predictable additional hit.

Then the tumultuous early 1990s gave way to the late 1990s, and a funny thing happened. People moved back. Housing prices recovered, and then soared beyond their previous highs. Not all of the wounds healed, of course, especially not for families and friends of those who had lost their lives. The process of rebound, no matter how tenacious or elastic, operates at a societal level. To recognize this process is not to minimize the devastating effects that earthquakes can have on a personal level. But as a whole, the region hardest hit by Northridge most certainly did recover: within a few short years one was hard-pressed to find any hint of the earlier devastation. Many damaged older structures, including the Northridge Meadows Apartments, where 16 people were killed, gave way to newer structures that were not only more modern but also far better designed to withstand future temblors.

Los Angeles's urban woes had not been erased, of course, by the start of the new millennium, but they had begun to appear more similar to those of other large U.S. cities. The relationship with the natural environment perhaps remains more precarious than that of most major cities. But reports of Los Angeles's demise were, it turns out, greatly exaggerated. In the 1990s it was the obvious conclusion, the *easy* conclusion: Los Angeles had begun the slow, inexorable process of self-destruction. The natural environment, relentlessly abused for decades, had begun to bite back. The wildlife had turned hostile (killer bears! killer bees!) . . . the mountainsides had turned hostile . . . the earth itself had turned hostile. Any sensible person would see the writing on the wall and leave, once and for all, as a half-million people in fact did in the years following the Northridge earthquake.

What gives? Had the California dream become too indelible to be erased by anything as minor as the most expensive earthquake disaster in U.S. history? Had the illusion become too resilient to evaporate in the face of the most obvious imaginable reality?

Were 15 million people really that stupid?

The easy answer, even when it is the satisfying answer, doesn't always turn out to be the right answer. Having now considered New Madrid, Charleston, San Francisco, and other stories, the alternative possibility looms as large as the nose on one's face: Could 15 million people simply be that resilient?

They don't call it Edge City for nothing. Consider for a moment the implications of this. Throughout history some large earthquakes have been very rude surprises—perhaps none so much as the 1755 quake, but other large U.S. earthquakes as well. Perhaps more than any other group, Californians of European descent never had a chance to imagine that their terra firma would remain reliably and faithfully firm. How could they, when the first European explorers were met with tales of earthquakes from Native Americans, and sometimes more than tales? One has to wonder if Friar Juan Crespi's assessment of an idyllic Santa Ana region wasn't made before the Gaspar de Portola expedition experienced a series of strong earthquakes in 1769 as they made their way north through what is now the city of angels.

California's history is inexorably intertwined with California's earthquakes. When James Marshall's discovery of gold at Sutter's Mill sparked California's first big wave of immigration, the northern San Andreas fault was, as it happens, nearing the end of its cycle. The nature of such cycles began to be appreciated only in the 1990s, as scientists considered data from large earthquakes worldwide and began to recognize a pattern. After a large earthquake occurs and its aftershocks die down, a region tends to go quiet for many decades because the stress in the system has dissipated—a bit like the toddler who sleeps like a lamb after a really, really big tantrum. Seismologists have a name for the quiet after the storm; we call it a stress shadow. After a while, stress starts to build again; the rate at which it builds typically depends on the plate tectonics forces in a given area. The shadow does not disappear with a bang, but erodes gradually. And, like the cheerful toddler who starts to get tired again after her long nap, the region around a fault starts to whine. In the earth, this whining takes the form of an increasing rate of moderate earthquakes—not on the largest fault, which remains locked, but on smaller faults throughout the region.

Thus, as the dominant fault in a region approaches the end of its cycle, that region will experience an increased rate of moderate temblors, the largest of which can be destructive events in their own right. With the benefit of hindsight, we of course know what late 19th-century California settlers could not have imagined: that the northern San Andreas fault was approaching the end of its cycle. Thus the early settlers in northern California were greeted with gold, golden sunshine, golden poppies, golden hills . . . and earthquakes. As the stress shadow around the San Andreas fault eroded, moderate earthquakes popped off with increasingly gay abandon. Felt earthquakes occurred, on average, every couple of months in and around San Francisco—a far greater rate of perceptible temblors than during the late 20th century.

Earthquakes were also a part of life for early European settlers in southern California. Southern California was sparsely populated when the 1857 Fort Tejon temblor tore along much of the southern San Andreas fault, but the region was rocked by several large and damaging earthquakes throughout the 20th century as well: Long Beach in 1933, Kern County in 1952, San Fernando in 1971.

As early as the late 1800s, one senses a certain defensiveness about earthquakes from Californians. In his 1872 book *Golden State*, Guy McClellan points out that "indeed, compared with the earthquakes of other times and countries, California's temblors are but gentle oscillations."[5] But in the less California-centric 1890 publication *Great Disasters and Horrors in the World's History*, A. H. Godbey writes, "For frequency of shocks, and total damage in consequence, the Pacific States far exceed the rest of the country. Their position with active volcanic regions in Oregon and Washington and Lower California, renders them peculiarly liable to such disturbances. Within the years 1872–1885, inclusive, there were registered seventy-five earthquakes in New England, sixty-six in the Atlantic States, seventy-five in the Mississippi Valley, and two hundred and thirty seven in the Pacific States."[6]

In addition to the sentiments expressed by historians and the public, one can look to the earth science community to gauge their assessment. Where earthquakes and volcanoes are concerned, scientists vote with their feet. That is to say, as a rule, scientists care deeply about their science, and don't want to study it from afar. They want to be where the action is. The Seismological Society of America was founded in the San Francisco Bay area, is still headquartered there, and today has an impressive 28 percent of its membership living in California. Even in 1906, the San Francisco region boasted more than its fair share of eminent geologists, men including Andrew Lawson and Harry Fielding Reid. They lived in California because, even before scientists understood faults, they understood that California is earthquake country.

That these pioneers of earth sciences chose to move to California, ignoring possible peril to life and limb, is not surprising. There is no question that scientists are a nutty breed. But what about everybody else? The millions who moved to California not because of earthquakes, but in spite of them?

Edge City. Los Angeles—indeed, California as a whole—did not wake up one morning to find itself unexpectedly on the edge. It was born that way. Like the generations of pioneers before them, early settlers arrived in the

Golden State well aware of the hazards of life in the Wild West. Later settlers arrived with their eyes open as well. The hazards posed by the natural environment have always been on display, spectacularly so, in California. Raging floodwaters washed away homes near the Los Angeles River in the great flood of 1938, and created lakefront property in downtown Los Angeles during the great flood of 1955. The disastrously wet winter of 1969 buried foothill homes in rivers of mud as massive debris flows occurred on hillsides throughout greater Los Angeles. The El Niño-driven storms of 1983 wreaked coastal havoc to the tune of almost $1.5 billion.

Walden Pond on LSD—that was Mike Davis's description of a city whose seemingly idyllic natural setting had turned out to have an unexpected, persistently, and sometimes malignantly psychedelic quality. The siren's call had swelled the population of greater Los Angeles to some 15 million souls, but, as the early 1990s made clear, it was time to pay the piper. The illusion had been shattered, once and for all.

Except that there was not, and never had been, anything to shatter. If anything, the *absence* of illusion had left its mark on the collective psyche of generations of Californians. Walden Pond? California was never Walden Pond—not from the days of the Portola expedition. Walden pond on LSD? More like Disneyland on steroids: life in California involves a lot of fun and a lot of thrills, the latter of a more potentially life-threatening nature than mechanical pirates. Nobody pays the price of admission at Disneyland expecting a day of quiet reflection communing with nature; nobody over the age of eight arrives in California with this expectation either. Those who do arrive have packed a sense of adventure along with the sunscreen. After the firestorm of 2003 consumed virtually the entire hamlet of Cuyamaca in the hills east of San Diego, residents Mona and Ivan Hecksher summed it up in a nutshell when asked by *Los Angeles Times* reporter John Balzar if they planned to stay and rebuild. "Of course," Mona said, "This is the very best place." Her husband shrugged and added, "Only crazy people live here."[7] Then he smiled.

Southern California: Land of Nuts and Flakes. Here again an answer that fits neatly with preconceptions. But can we believe that 15 million people are (truly) crazy any easier than we can believe that 15 million people are stupid? Objective evidence suggests otherwise. Decades of ongoing lunacy have led to a natural selection process as inexorable as evolution. Many people stay in any area simply because they were born there, but even today California has more than its share of immigrants—refugees from other countries as well as from parts of the United States with harsh winters and/or moribund economies. Between the beginning of 2000 and the end of 2003, fully a million newcomers—men, women, children, babies—took up residence in southern California. Assuming that many of these new arrivals were newborns, one shudders to imagine what area kindergartens will be like in a few years.

By the time they are into adulthood, many of California's native sons and daughters have made a choice to stay. Some people may lack the economic means to relocate easily, but people can leave—as evidenced by the fact that people *have* left, at times seemingly in droves. Most adult Californians are thus Californians by choice, in spite of the myriad and often dramatic hazards that are part and parcel of California life. This selection process has created a population that is willing to accept the bad along with the good, a population that is resilient . . . a population with a sense of humor. When an M3.7 earthquake struck near Simi Valley in the midst of the 2003 firestorm, a headline on the KFWB Web site read, "Earthquake Strikes Simi Valley: Can Locusts Be Next?"[8] And when an itinerant Class A baseball team moved to the city of Rancho Cucamonga in 1993, in the very shadow of the San Andreas fault, what else would they call themselves but The Quakes? Team mascots Tremor and Aftershock entertain fans at a lovely stadium known as (what else?) The Epicenter.

Evolutionary forces have perhaps selected for another trait as well: a realistic mind-set. Not in a general sense, of course, but rather a very specific one. The point has been made many times, in books as well as across dinner tables, that California by no means has a monopoly on disasters. Throughout U.S. history, some of the costliest disasters, in terms of both dollars and lives, have been hurricanes along either the Atlantic or the Gulf Coast. Even without hurricane-force winds, fierce East Coast storms, including the classic Nor'easters, can take a heavy toll. Much of the Midwest, meanwhile, comes under assault by tornadoes on an all-too-frequent basis. Not to mention the fact that virtually no place on earth is immune from earthquake hazards.

Thus we are left with one of the realities of life on a dynamic planet: virtually no place on earth can be deemed safe from all natural disasters. Considering, dispassionately, the toll taken by natural disasters of various stripes, however, another reality emerges: as a rule, human beings are scared of the wrong things. One's odds of dying in an earthquake, hurricane, flash flood, avalanche, or other natural disaster are very, very small—even in California, even recognizing that, sooner or later, a Big One will strike the heart of an urban area and claim a death toll numbered in the thousands.

Certain fears have a primal quality to them, quite apart from the realm of statistical analysis. To be overtaken by a firestorm . . . to be wakened out of a deep sleep by the sounds and fury of a major earthquake . . . such is the stuff that nightmares are made of. These sorts of cataclysmic disasters tear at the heart of the carefully constructed illusions that get us through the day: that we inhabit a basically sound, predictable, *rational* world. If you can't count on the terra firma to stay firm, what can you count on?

Viewed through the dispassionate lens of statistics, the peril of cataclysmic disasters is really very small in the larger scheme of things. In an average year, about 40,000 Americans die in automobile accidents. Over a given 80-year life span, that adds up to over 3 million body bags. With about 300 million people in the country, these numbers reveal that, over any one person's lifetime, about 1 percent of his or her fellow citizens will be killed in automobile accidents. Put another way, any one person has one chance in 100 (give or take) of dying in an automobile accident. Yet, as a rule, cars don't scare us; they are too much a part of day-to-day life, too seemingly harmless to be scary. The cumulative hazard posed by automobiles can be so invisible that thousands of drivers fail to take even the most basic precautions: wearing seat belts, driving no faster than the flow of traffic, slowing down in the rain.

But taking dispassion one step further and considering how people really tend to die in modern industrialized nations, it becomes apparent that what we have to fear is not fear itself, but rather ourselves. A 2000 study by Majid Ezzati and Alan Lopen concluded that a staggering 12 percent of global adult deaths were due to tobacco-related causes. These deaths occur primarily among the elderly, so tobacco does not shorten global lifespan by a commensurate amount, but still, the numbers are enormous. The toll taken by obesity and/or lack of fitness is harder to assess, but recent studies have linked these prevalent conditions to an increased risk of any number of potentially lethal health problems, including heart disease and cancer.

And there you have it. Earthquakes don't kill people; buildings only rarely kill people—tobacco and french fries kill people.

It would be difficult to assess the collective level of dispassion—or appreciation of the statistical issue of relative risk—of any group of people. But Californians seem to have more appreciation than the average bear for the fact that no place on earth is truly safe from natural disasters. They also arguably have more appreciation than most for the perils of tobacco and french fries. According to a 2002 study by the Centers for Disease Control, 17.2 percent of California adults smoked as of the year 2000—a lower incidence than every other state except for Utah, and well below the national average of 23 percent.

Considering obesity statistics for the year 2000, one finds California in the large group of states where the rate is 15–19 percent. Only Colorado fell below this range, while 22 states fell in the greater than 20 percent bracket. In this respect Californians appear to mirror larger societal trends, although they still do better than the national average.

It is debatable, of course, whether or not Californians have a better sense of relative risk than do Americans in other states. (They have been accused of having a better—or at least *different*—sense of all sorts of other things.) And as previous chapters have illustrated, Californians certainly have no monopoly on resilience. Like human beings everywhere, Californians draw on deep reservoirs of elasticity to cope with the disasters—natural and otherwise—that come their way. But in the Golden State the theme of resilience plays loudly and often; without question this leads to a natural selection process that shapes the collective psyche of those who stick around.

Only crazy people live here.

If you choose to live on the edge, you have to be prepared to hang on for the ride.

12

Earthquakes as Urban Renewal?

Even, therefore, if San Francisco was visited by a calamitous earthquake, its progressive career as a city would be but temporarily interrupted, and though real estate and other values might suffer from an immediate panic, they would quickly recover again.

— *San Francisco Real Estate Circular* (October 1868)

Whether or not the reader finds it convincing, by now the thesis of this book is clear. At least throughout recent history, earthquakes have taken a temporarily heavy toll in some areas, devastating cities, claiming lives, and shaking faith. Yet taking a step back to consider the longer-term impact, one finds that, almost without exception in recent historic times, cities and societies rebound with elasticity to mirror the earth itself. Elastic rebound. These two words represent not only the single most fundamental tenet of earthquake theory but also the most apt metaphor to describe societal response to even the most catastrophic seismic events.

As previous chapters have illustrated, mankind's capacity for elastic rebound is largely a reflection of *man's* capacity for elastic rebound. Recall the challenge to Voltaire, "Alas, times and men are like each other and will always be like each other."[1] These words might have been penned in the context of matters of philosophy: How do we make sense of our existence and our place in the universe? But at the end of the day, most days have not concerned themselves with philosophy, and politics is left to the politicians. At the end of the day, people are people. When a devastating earthquake strikes, perhaps the complex superstructure of society crumbles along with the buildings. When elaborate social and political façades are stripped away, perhaps the finer inclinations of the individual are not changed but rather showcased.

It's a nice thought, at any rate. Whether or not it explains the predilection for resiliency and compassion following disasters is open to debate, but a consideration of history, as outlined in the previous chapters, suggests that the predilection is real—whatever the cause. This remarkable human capacity for rebound is clearly a critical factor mitigating the overall societal impact of earthquakes and other natural disasters.

But resiliency and compassion alone cannot hope to rebuild modern cities following a major loss of life and property: recovery requires resources. Having considered important individual earthquakes at some length, we now turn to a general consideration of the economics of earthquakes.

Considered dispassionately, one can make the argument that earthquakes invariably become a catalyst for urban renewal. Once built, any structure possesses a tremendous amount of inertia: the supposedly temporary Quonset hut that finds itself in use 50 years later, the masonry building in Los Angeles that is too expensive to retrofit and too valuable to tear down. In entire cities in countries such as India and Pakistan, many people cannot afford basic necessities of life, let alone the costs of making their dwellings earthquake-safe.

Building codes provide no guarantee that buildings will come through a major earthquake unscathed, but they surely do increase the odds of survival—for the occupants as well as the structure. Enforcement of building codes represents another layer of challenge, one that has proven especially vexing in countries such as Turkey, which historically lacked a strong tradition of enforced building inspections. These limitations notwithstanding, by far the best chance of building an earthquake-resistant city is to build the buildings right in the first place, according to strict and strictly enforced building codes.

Earthquake resistance comes at a price, of course. In the United States, the incremental cost has been estimated at 1–3 percent of total building costs—painful but rarely prohibitive. In other parts of the world the calculation can be more difficult, the incremental costs higher, and the bottom line more difficult to absorb (see sidebar 12.1). For large-scale government building projects, such as a recent project to build new schools in India, the incremental

SIDEBAR 12.1

What is the cost of earthquake resistance? It can be a surprisingly difficult question to answer. The answer clearly depends on the nature of the
building. For the new modern, well-built structure in industrialized nations, practicing engineers cite an incremental cost of perhaps 2–3 percent at most: "less than the cost of the carpet," they sometimes say. In
developing countries, where buildings are constructed more cheaply, the
incremental cost of earthquake resistance can be higher. A figure of 10–20
percent might be reasonable, although this remains a rough estimate of
a quantity that is highly variable. (If it costs nothing to erect a home
from existing stones, the incremental cost of earthquake resistance can
be infinite.) The cost of resistance also depends greatly on performance
expectations: whether a building is supposed to survive an earthquake
unscathed or simply not collapse on its inhabitants (even though it
might be damaged).

cost might be somewhat higher than in the United States. But if a country is
desperately lacking schools for all of its children, it is no small matter to
choose to build 5 or 10 percent fewer schools. And consider the subsistence
farmer in Turkey or China, whose house consists of stones glued together
with cheap mortar. The cost of such a structure, or of rebuilding a structure
following a major earthquake, is very low; even a modest incremental cost
can essentially be astronomical compared with the resources of the homeowner. Considering the equation in a broader context, if you aren't sure
where your next meal is coming from, insurance against earthquake damage
is a luxury you cannot afford.

Now consider the same equation in the immediate aftermath of a devastating earthquake. The potential severity of earthquake damage will never be
appreciated as deeply as it is right after severe damage has happened. With
dramatic imagery deeply and recently etched in the collective psyche, individuals and governments alike find themselves motivated as never before.
Individual homeowners take steps to shore up their houses, formerly reluctant government agencies have a newfound willingness to invest in mitigation, monitoring, and research programs.

When the damaging earthquake occurs in a country with limited resources, a formerly reluctant *world* can have a newfound willingness to contribute money and expertise. Assistance comes from a range of sources, not
only foreign governments but also agencies such as the World Bank, as well
as national and international charity organizations.

In a sense, our collective humanity serves as an enormous insurance
net—one that will bank resources in innumerable places and dispense resources when and where they are critically needed. A major disaster effectively
draws on premiums that have been paid over time, potentially by individuals and organizations throughout the world. Unlike conventional insurance,
this larger effective insurance net does not distribute the premiums according to risk, but more typically according to ability to pay. When the chips
are down, even capitalistic industrialized nations find a bit of socialism in
their hearts.

If traditional insurance companies have done their homework, conventional insurance is a bit better than a zero-sum game for any one region: the
benefits paid out will, more or less, equal the premiums paid in, minus the
company's profit. Still, the benefits will be paid out after the earthquake (or
other disaster) strikes.

A major earthquake focuses resources. With conventional insurance, resources are focused in time: premiums are paid in over the long term, and
paid out after disaster strikes. With effective societal insurance, resources
are focused in a regional sense: collective banked resources are tapped after
disaster strikes. Thus do we arrive at the paradoxical conclusion: a major
earthquake can in the long term represent an economic boon for the hardest-
hit areas.

Insured losses are covered very directly, generating a straightforward paper
trail: payments go directly to policyholders and thence directly to contractors, bricklayers, Home Depot, and so forth. Uninsured losses in the United
States are compensated to some extent by low-interest loans from the Federal Emergency Management Agency, providing money that goes in the same
directions as insurance dollars. Following an earthquake as large and damaging as the 1994 Northridge event, local contractors can find themselves with
more business than they know what to do with. The situation was not too
much different following the Charleston earthquake in 1886, before earthquake insurance or FEMA existed. For the most part, people find a way to fix
damaged property. Public property, meanwhile, is repaired with public dol-

lars. This money also flows in a number of directions: contractors and engineering companies, and so forth.

The safety net is not, of course, perfect. Uninsured homeowners find themselves with expensive repairs for which low-interest government loans are available—but the loans are not gifts. Renters in low-income areas can find themselves with the short end of the stick as well. In working-class neighborhoods in Watsonville, California, after the 1989 Loma Prieta earthquake, many renters found themselves displaced from low-income units that would not be rebuilt as low-income units.

In poorer countries, even an influx of global assistance may not be enough. The devastating 1988 earthquake in the Republic of Armenia dealt an especially severe blow to a number of cities, including Gyumri (formerly Leninakan). This earthquake represented something of a worst-case scenario for earthquake relief. Both the region and the former Soviet Union were preoccupied at the time with tumultuous political and economic situations. A newly independent country following the dissolution of the Soviet Union— also a country that faces harsh winters—Armenia had its share of economic challenges without the added injury cased by a major earthquake. The earthquake, then, added the proverbial insult to injury. Over a decade later, recovery is still slow in reaching the hardest hit cities. In Gyumri, thousands of people were still living in their "temporary" *domiks*: dwellings hastily constructed using converted cargo containers.

Considering even the most deadly earthquakes in recent historic times, however, the Armenian event emerges as the exception, not the rule. The deadliest earthquake in history, the 1976 Tangshan, China, event, claimed a staggering toll in a swath across northeastern China. Published death toll estimates have varied widely; at the time the earthquake struck, the government of China was not forthcoming with such information. Current estimates range from 250,000 to as high as 750,000 people. By some estimates over 90 percent of residential buildings and almost 80 percent of commercial buildings were completely destroyed.

The tight controls imposed by the Chinese government also limited the dissemination of information about the recovery. By most accounts, however, the city's industrial base had been rebuilt within 10–20 years. With individual fortitude and an influx of resources from an enormous and centralized government, even Tangshan—now "The Brave City of China"—was able to rebound. Interestingly, the city has chosen not to forget its tragic legacy. Author and Cabrillo College history professor Sandy Lydon leads groups of Californians to Tangshan and points out the seven damaged sites, including a library with several pancaked floors, that have been left unrepaired, memorials to both the dead and the heroic efforts of survivors.

When a major earthquake channels money and other resources to an area, several things happen. Most obviously, property and infrastructure get rebuilt. But the effects go beyond the obvious. At no time in history are people and governments as aware of seismic hazard, and as willing to do something about it, as they are in the immediate aftermath of a devastating earthquake. In recent times in the United States, "doing something" has included spending federal dollars on hazard mitigation research. Major earthquakes propel earthquake science forward not only by the bounty of new data they provide but also, invariably, by a bounty of research dollars as well. Consider the state-of-the-art seismic network in southern California, which in all likelihood would not exist if not for the Northridge earthquake. The devastating 1995 Kobe earthquake, which, ironically, occurred a year to the day after Northridge, inspired similar stepped-up earthquake monitoring efforts in Japan. Within the United States as a whole, earthquake monitoring became a much higher priority after the 1964 Good Friday earthquake in Alaska.

In industrialized countries, rebuilding efforts also include a newfound level of respect for risk mitigation. Often this respect is implicit, a consequence of the fact that rebuilding will be governed by current building codes, not whatever codes, if any, were in place at the time a structure was originally built. At other times the respect is based on sensibilities heightened by dramatic recent events. When an earthquake causes widespread damage to masonry chimneys, many homeowners opt for new modular chimneys, even if current codes do not outright forbid the construction of masonry ones.

In both the public and the private arenas, earthquakes also motivate spending on structures that were not damaged by the latest temblor—but clearly might be impacted by the next one. Consider the stepped-up program of freeway overpass retrofitting in California following the Northridge earthquake.

"A new and more beautiful, more finished city had sprang up in the ruins of the old."[2] It was as true of Yokohama, Japan, in 1923 as it was of Charleston, South Carolina, in 1886, and of San Francisco in 1906. Earthquakes are the earth's way of correcting the mistakes made by human beings: If a building isn't well suited to this environment, down it goes. If that city lacks the infrastructure to combat a large-scale conflagration, down it goes. The process is nothing less than Darwinian. If you want to build your house or your city on the edge, figuratively or otherwise, you better plan accordingly. And in fact, people do—if not before the fact, as they should, then at least after the fact.

Earthquakes as urban renewal. An odd assessment, one that expresses ghoulishness and optimism in equal measure. But if we accept this viewpoint, where does it lead? Perhaps to an optimistic flavor of fatalism: no sense worrying about earthquakes, history proves that society will recover (see sidebar 12.2). Indeed, this sentiment has been expressed in writing at least once. In the aftermath of a severe 1868 temblor on the Hayward fault—for a few decades, *The* Great San Francisco earthquake—an article in the *San Francisco Real Estate Circular* concluded, "Even, therefore, if San Francisco was visited by a calamitous earthquake, its progressive career as a city would be but temporarily interrupted, and though real estate and other values might suffer from an immediate panic, they would quickly recover again."[3] That the sentiment proved true, more or less, does not gainsay its fundamental wrongheadedness. Large earthquakes can exact a terrible price from both individuals and society as a whole, notwithstanding deep wellsprings of resilience. When a 2003 earthquake leveled the ancient, historic city of Bam, Iran, 30,000 lives were lost in little more than a heartbeat. Some losses cannot be tallied with dispassion, and the potential of massive, widespread devastation and loss of life from earthquakes far exceeds that from other types of natural disasters.

When the earth unleashes its harshest furies, the immediate price is, perhaps more than anything else, psychological. In *Great Disasters and Horrors in the World's History*, A. H. Godbey writes:

> Man's social arrangements are calculated upon a supposition of the earth's stability: and when he finds himself the victim of misplaced confidence, there is neither courage nor spirit nor reason left in him. Numerous are the cases where men have been rendered insane by such convulsions.
>
> To the ravage of the hurricane, the roar of the storm, the surge of the sea, the rush of the flood, one becomes in measure accustomed, and in the moment of danger may take precautions for personal safety. But in the case of earthquakes the reverse is the rule; none dread them more than those who know them best.[4]

Godbey's discourse continues:

The sensation of utter powerlessness is so overwhelming, that amid the crash of falling houses, the cries of entombed victims, the shrieks of flying multitudes, the rumblings in the earth beneath, and the trembling of the soil like that of a steed in the presence of a lion, the boldest and bravest can but sit with bowed head, in silent, motionless despair, awaiting whatever fate a grim capricious chance can determine.[5]

Mr. Godbey can perhaps be accused of hyperbole, but there is little question that abrupt, unheralded, severe ground shaking elicits a singular immediate terror, even today.

Dr. Pradesh Pande, a lead geologist with the Geological Survey of India, has led several teams of scientists in the collection of detailed damage reports following large earthquakes in India. Pande observes a systematic difference between individuals who have experienced moderately damaging shaking and those who have experienced shaking levels approaching the worst our planet has to offer. Among the former group one finds, as a rule, individuals who are ready, willing, and able to describe their experiences in some detail. Among the latter group one finds, as a rule, people who have to struggle even to find their voices, in the first few days especially.

Of those who do experience such terrors firsthand, many survive. But, of course, in developing nations especially, large earthquakes come at an especially steep price when tallied in terms of human lives. The 2001 Bhuj earthquake in western India claimed nearly 14,000 lives; the 1976 Tangshan earthquake in China claimed more than 20 times more. Only a metaphorical handful of people have lost their lives in earthquakes in the United States, but experts who consider such things do not expect this luck to continue. An M7.5 temblor in the heart of Los Angeles, or perhaps an M6.5 temblor in the heart of New York City, could easily claim a death toll numbering well into the thousands. Such events would leave the Northridge earthquake in the dust, so to speak, on the list of most costly U.S. disasters.

Dispassionate consideration of economic issues aside, an important point remains: earthquake losses cannot possibly be tallied in full on any spreadsheet. When the loss of human life hits close to home, the wound never heals. This is as true if the number of such wounds is 40 or 40,000. When the numbers reach into the tens of thousands, the impact on society can be profound as well: so many people lost, so many children, so many parents. Iran will recover from the devastating 2003 Bam earthquake; the city that rises

SIDEBAR 12.2

Entire books have been written on the theme that earthquakes are powerful agents for social change—quite a different theme from that of this book. Indeed, in some cases, the years following earthquakes can be characterized by a different metaphor: not rebound but rather disintegration, if not outright implosion. Paul Reynolds, a BBC correspondent, dubbed such events *political earthquakes*: Armenia in 1988, Nicaragua in 1972; some would add Tangshan in 1976. Clearly, as the Armenia earthquake illustrated, some hard-hit cities do not recover for a very long time, if ever. Yet closer inspection reveals a commonality between most temblors with such (apparently) far-reaching reverberations: the seeds of political instability had been planted years, if not decades, before the ground shook. In Nicaragua, Somoza's ineffective rescue and rebuilding effort brought to the world's attention the nature of his dictatorship, but even so the Sandinista uprising did not take place until 1979. In China, the Tangshan disaster was viewed as a bad omen, and left-wing leaders capitalized on the tumult to flex their political muscle. Yet Premier Zhou Enlai had died earlier that same year, and Mao Zedong was ill in bed by the time the earthquake struck. The winds of change were clearly blowing: the earthquake might perhaps have acted as a catalyst, but it was not the cause of the political upheaval that followed. Still, such examples illustrate an important point when we ponder the world's responsibility and interests in future earthquake disasters: where a country is strapped for resources or already destabilized by economic or political forces, a devastating earthquake can represent a catalyst whose effects will be difficult to predict or control. Consider the possibility of a truly devastating earthquake in Iran or Pakistan, for example, at a time when powerful factions threaten the country's political fabric. Catalysts may not cause reactions, but they certainly can accelerate reactions. When the experiment involves a mix of politics and earthquakes, modern society may, as a rule, rebound well, but the potential for a runaway reaction—for political meltdown—cannot be disregarded. This potential will, moreover, likely be highest in precisely those countries one worries about most: the ones that already impress us as unstable and volatile in ways that can keep us up at night.

from the ashes will almost certainly be more modern, better able to withstand the large earthquakes in Iran's future. The terrible losses may further motivate risk mitigation and modernization in other parts of the country. But the toll on the city and its population remains staggering: 30,000 lives lost, a richly historic city gone. Around the rim of the Indian Ocean, the loss of life was as much as ten times larger still in the 2004 Sumatra disaster. And the hardest realization of all: *this price need not have been paid.* Earthquake hazard is inevitable on a dynamic planet; earthquake *risk* is not.

Earthquake research efforts in the United States and elsewhere now focus on hazard assessment and risk mitigation because this is where one finds real bang for one's research buck. Dollars spent on such research lead to an improved understanding of the shaking during large earthquakes; with sensible knowledge transfer programs this leads in turn to better building codes and better buildings. Even if scientists could predict large earthquakes, the fact remains: buildings and infrastructure would still have to be built properly. This task is much easier in places that have, or can afford to build, homes of wood, but masonry buildings can be built earthquake-safe as well, in some cases with very low-tech measures (for example, rebar reinforcement of concrete.) Other types of masonry construction pose more of a challenge, for example, in parts of the world where homes are often little more than piles of rocks held together by not very much.

It is not realistic to hope that earthquake risk will someday be mitigated out of existence. Even if we could predict earthquakes, we cannot stop them. And even if we design every structure to withstand earthquakes, there will still inevitably be fatalities: in structures that were not built to code or in which the contents were not secured, or on freeways that instantaneously turn into roller-coaster rides. However, the contrast between the 1994 M6.7 Northridge earthquake and the 2003 M6.6 Bam earthquake effectively illustrates the degree to which risk can be mitigated: about 50 lives lost in the former, over 40,000 in the latter.

One might wonder, even, if the societal capacity for rebound doesn't sometimes work against the societal potential for risk mitigation. In the central United States, respected seismologist Seth Stein has in fact made the argument that the potential losses from future large New Madrid earthquakes do not warrant the cost of designing buildings to withstand them. Indeed, the 1811–1812 New Madrid earthquakes were barely a speed bump in the path of westward expansion, and the Midwest will surely recover from whatever the New Madrid zone dishes up next. If the New Madrid or Charleston or San

Francisco earthquakes had marked the death knell of their respective locales, it seems unlikely that anyone would argue against the need to prepare for the eventual recurrences of these temblors. That we can absorb the losses does not, however, imply that we *should* absorb them. The cost of proper initial design is many times cheaper than the cost of retrofitting later, and it is also a mere pittance compared to the cost of significant earthquake damage.

The question of earthquake losses—both lives and dollars—can now be addressed quantitatively, and used to make reliable predictions for losses from earthquakes still to come. Having considered the past at some length, we are now poised at last to consider the future.

13

Demonic Demographics

The end of the human race will be that it will eventually die of
civilization.
> —Ralph Waldo Emerson, *The Essential Writings of Ralph*
> *Waldo Emerson* (New York: Modern Library, 2000)

Between 100 B.C. and about A.D. 1600, global populations doubled from
around 300 million to more than 600 million people (figure 13.1a). The second
doubling in world population was very much faster. It occurred between
1600 and 1800, when improvements in medicine and living conditions
resulted in a dramatic reduction in mortality rates. The third doubling in
world population, from 1.2 billion to 2.5 billion was faster still, in the 150 years
following 1800—in spite of the occurrence of two world wars. The fourth
doubling occurred in less than 40 years, bringing global population in 1990
to more than 5 billion. Although the rate of population increase has slowed,
a fifth, and perhaps final, doubling of population in the next 50 years is projected
by the United Nations, bringing projected world populations in 2050
to somewhere between 7.6 and 10.6 billion.

In the past 200 years, then, we have increased the number of people on
our planet by a factor of 10. We might conclude from this that whatever the
impact of earthquakes on human history, it may be very different from what

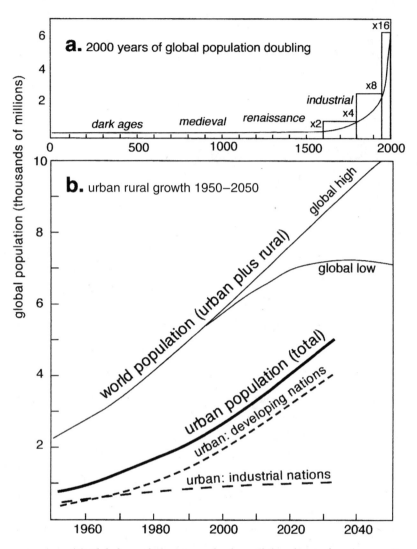

FIGURE 13.1. (a) Global populations grew slowly until the advent of modern medicine controlled diseases in the world's cities. Since 1600 world populations have doubled, redoubled, and doubled again. They continue to rise, with growth occurring mainly in cities. (b) According to U.N. predictions, by 2050, half of the world will live in cities and almost all of the world's urban future growth will occur in the developing nations. In 2000 half of the world's urban dwellers lived in cities with fewer than 500,000 people, where growth is rapid, and 4 percent lived in megacities (10 million), where growth is slower.

we may expect in our future. Neither the rates nor the distributions of large earthquakes have changed appreciably in millions of years, but the risk has grown simply because the sizes of the targets are now so much greater than ever before.

Of particular note for the present discussion, the recent tenfold increase in human populations has occurred largely in cities. During the Middle Ages the rural population outnumbered the urban population by about 100 to 1, a ratio that reflected the high risk of communicable disease in cities. Cities were essentially places for the excess rural population to move to, and to die young. Advances in medicine upset the natural mortality of cities that had checked their growth, and urban populations have grown steadily since 1600. Once mainly rural dwellers, we have increasingly become urbanites with a minority living in the countryside. United Nations projections anticipate that half the entire world will live in cities by 2050 (figure 13.1b).

This concerns those who study earthquakes for several reasons. Catastrophic earthquakes have a habit of returning at intervals on the order of 100–500 years, and smaller ones, more frequently. The overall human target is thus now ten times larger than it was when the 1755 Lisbon and 1811–1812 New Madrid earthquakes struck. Worse yet, more than half of the world's largest cities lie preferentially on plate boundaries, where great earthquakes not only are expected, but have in earlier times destroyed the villages that have now grown to be megacities.

The human target for earthquakes is thus not only larger but also preferentially concentrated in all the wrong places. Belts of earthquake-damaged cities follow the Alpine Himalayan collision zone from Europe to Indonesia. They encircle the Pacific Rim, and they encircle the Caribbean. On an increasingly urbanized planet, this unfortunate concentration implies an ever-increasing risk (figure 13.2).

Worse still, the vulnerability of the human target will almost surely increase dramatically over the next few decades. A staggering 2 billion or so new souls are projected to arrive on the planet over the next 20 years. During this time, the population of cities in developing nations is projected to grow by, give or take just a bit, 2 billion. This apparent paradox has a twofold explanation: first, the fact that populations of the industrialized and seismically safest countries have stabilized and, second, the fact that migration patterns will continue to funnel people into cities such as New Delhi, Kathmandu, Caracas, and Mexico City. The largest current growth of the world's urban population occurs in cities with populations of less than 500,000 in

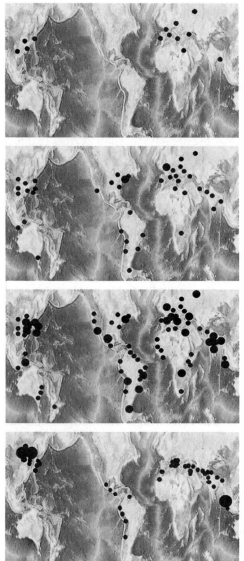

Pre-1600
city populations
less than 1 million

1950
27 supercities (2 million)
2 megacities (8 million)

2000
325 cities (1 million)
140 supercities (2 million)
28 megacities (8 million)

earthquakes
(1000-2004 AD)

• 10,000-100,000 dead
● more than 250,000 dead

FIGURE 13.2. Growth of cities since 1600 compared to the numbers of people killed by catastrophic earthquakes (more than 10,000 dead) in the past millennium. Note that many urban agglomerations have grown where earthquakes have historically caused large loss of life.

the developing nations, cities that will eventually become megacities in two to three decades, given current growth rates. The continued growth of cities in the developing nations does more than put more people in harm's way: with increasing land pressures, traditional building styles often give way to bigger residential structures, many of them constructed from carelessly assembled concrete. The 2001 Bhuj, India, earthquake killed people in the city of Ahmedabad, some 150 miles from Bhuj, largely because poorly assembled concrete apartment units collapsed. The Izmit and Duzce earthquakes in Turkey in 1999 resulted in a similar number of fatalities for very similar reasons.

As previous chapters have discussed, seismologists are careful to differentiate between hazard, the exposure of a region to earthquakes, and risk, which reflects the peril of structures, given an area's natural earthquake hazard. The distribution of earthquake hazard evolves only on the timescale of plate tectonics, which is to say that, as far as humans are concerned, hazard does not change. Risk, however, is poised to change enormously over the next few decades—and not in the right direction.

City Blocks as Sitting Ducks: Do We Have Any Choice in Locating Cities?

Every city on the planet represents a potential target at which the planet might aim an earthquake. Where are these cities located, and how do they change in time? A curious human behavior pattern, well known to geographers, is that early in the course of civilization, a hierarchy of settlements developed on every continent. Individual dwellings, sometimes grouped into hamlets, formed the base of the hierarchy, and were interspersed with larger villages where people shared their produce on market days. Villages were in turn subservient to a smaller number of towns, the largest of which acted as administrative or trade centers or as fortified enclaves. As time progressed, each settlement grew in size, but the relationships between them changed little. The hamlets became villages; the villages, towns; and the towns, cities. In recent times, as each settlement has doubled and redoubled in size, the largest cities have become what the United Nations describes as *supercities*, with populations of at least 2 million, or *megacities*, with populations exceeding 8 million. A comparison between Roman France and present-day France shows exactly the same hierarchy of settlements—they have merely grown in size. The ratios may not have changed, but the planet's largest modern cities represent an entirely new experiment for mankind: cities are now the size of classical nations.

From this persistent hierarchy we conclude there must be something special about the original distribution of cities. Once established, city locations are hard to change. The locations of cities appear to be essentially preordained. Our cities owe their existence to some special early attribute: a pivotal location with access routes through mountain passes, a river crossing, a harbor. Not of few of these geographical attributes are imposed on the earth's surface by plate tectonic processes. As illustrated in figure 13.2, many, but not all, of the known geographic triggers for cities are plate boundaries. Which is to say that nobody (except perhaps for daft earth scientists) gravitates to active plate boundaries *because* they are active plate boundaries. Rather, plate boundaries coincide with natural amenities—most notably coastlines—to which people gravitate, often unaware they are putting themselves in harm's way.

Digressions on the Nurture and Abandonment of Cities

Most cities thus develop because of the geographical amenities offered by their locations. Few artificially "grown" cities can be found on a map of the world, but there are indeed a few that were settled in spite of, not because of, their setting. Calcutta was founded in the late 17th century on an unhealthy mud bank on the Hooghly River to funnel trade out of an expanding empire. Canberra, Islamabad, and Brazilia were all formed relatively recently, and artificially encouraged to expand, in each case driven by unique national conditions. Although an ancient city, Tehran was in 1788 designated Iran's new capital following the destruction of the ancient capital, Isfahan, by an earthquake. Most North American cities are relatively recent, but even here their hierarchical development can be discerned. Artificially stimulated cities in the United States include Washington, founded by a nation in search of a capital, and Las Vegas, an artificial oasis in the desert founded by casino owners in search of gamblers.

Do cities die? They can. Many ghost towns exist because their reason for existence suddenly vanished. Ghost towns exist in the United States because the gold or other minerals that drew their inhabitants became exhausted. In the ancient world, cities died when their water supplies dried up. Cities in deserts along the Silk Road vanished when glaciers no longer sustained the rivers that permitted their existence. City populations soon abandon their town if a life-nourishing river finds a new course. Coastal cities have died when they have been submerged by the rising sea. Cities near volcanoes have been buried along with their inhabitants, as happened in spectacular fashion at Herculaneum and Pompeii. Ports are abandoned when the sea no longer floods the harbor, or when a growing delta strands a trading post far inland.

Cities have also been abandoned because of disease, but not for long. Portuguese colonists abandoned the old city of Goa to form the new city of Panjim when in 1840 it became clear that to dwell long in Goa's unhygienic streets was to die. Now that disease is no longer an issue, populations have returned, and old and new Goa are as populated as before. The Black Death trimmed European populations by one third, causing a brief halt in the expansion of cities, but no actual extinctions. China has a similar history of diseases affecting city size but with no lasting impact. Although extreme natural forces can succeed in extinguishing cities altogether, history tells us that these cases are the exception rather than the rule.

Historical Catalogs: Hit Lists of Future Disaster?

The association between cities and historically damaging earthquakes is quite well documented. For the memory of an earthquake to survive, the details must be written down or they are soon forgotten. Where best to find historians but in the cities of the world, present and past? In fact, many earthquakes are named for the city they destroyed, or for the one that suffered the most damage. The tradition of naming earthquakes after the hardest-hit city continues. In some cases no date is necessary—the town and the disaster are synonymous in the context of earthquake discussions: San Francisco, Bam, ChiChi, Tangshan.

The somewhat optimistic thesis of this book notwithstanding, one should not gloss over the fact that earthquake disasters can be—and *have been*—horrific (figure 13.3). Humans have already witnessed two earthquakes that extinguished on the order of a half-million lives, both in China (1524 and 1976). We have not discussed either of these events at length in this book because both remain shrouded in enigma, the former because of its antiquity and the latter because of the political situation in China at the time the earthquake struck. To this day, estimates of the death toll of the Tangshan earthquake remain uncertain to a rather astounding factor of 3: credible estimates range from a low of 250,000 to as high as 750,000 because whole families were lost, along with city records, during the earthquake.

FIGURE 13.3. One of the world's worst urban earthquake disasters reduced Tang-shan, China (1976), to heaps of rubble. More than 250,000 people died in Tangshan, though some estimates suggest that this number may err on the low side by a factor of 2–3.

Our view of the 1923 Kanto earthquake disaster is much sharper, even though it occurred a full half-century before Tangshan. Information flowed out of Japan immediately—information and photographs that captured the death and destruction in the starkest possible terms. So gut-wrenching were some of the images—the remains of tens of thousands of lives extinguished by the Hongo Ward conflagration—that decorum prevented their distribution in the West. Or, rather, decorum delayed their distribution for the two decades that it took for Japan to evolve from an admired ally to a despised foe. That Tokyo and Yokohama rose from the ashes cannot negate the extent of the tragedy and suffering these (and other) cities experienced.

In more recent times, the miracle of modern satellite communications provides news agencies with the ability to beam horrific images of earthquake disasters from anywhere in the world to television sets in living rooms, very nearly instantaneously. In 1985 the world not only watched with sadness as images of collapsed high-rise buildings in Mexico City were beamed around the world but also cheered when, against all odds, infants were found alive in the rubble days after the powerful Michoacan earthquake had struck. Equally gripping images, far more often heart-wrenching than heart-warming, followed from other earthquakes at other times: Kobe, Japan, in 1995, Turkey and Taiwan in 1999, western India in 2001, and Sumatra and elsewhere around the Indian Ocean in 2004. Again, one cannot emphasize the point strongly enough: society's capacity for rebound notwithstanding, earthquakes take a terrible toll not only on property but also on lives.

The seismologist or engineer is very much inclined to view earthquake losses as preventable: we *can* do something to mitigate losses, therefore we *should*—we *must*—do something to mitigate losses. Certainly it is a very bitter pill to swallow for any individual who loses a loved one and comes to understand that the death could have been prevented. The special and tragic case of the 2004 Sumatra earthquake, where many lives could have been saved with very modest prior investments, is a bitter pill to swallow. But a difficult reality emerges in a complicated world: it costs money to make buildings earthquake-resistant, and these costs take money away from other needs, many of them also pressing. Repugnant as it may seem to consider a cold-blooded cost–benefit analysis of any equation that includes human lives, the argument has been made, usually implicitly but sometimes explicitly, that earthquake resistance is not worth the price. And so we proceed to consider earthquakes in similarly cold-blooded dollars-and-cents terms: What has been the cost, in terms of human life, of past earthquakes on our urban planet? What is the expected cost of earthquakes yet to come?

Tallying the costs of past earthquakes, in terms of either deaths or damage, is a surprisingly difficult endeavor. Numerous catalogs of past disasters exist. Some are global and some are national. None are complete, and few are accurate. They contain exaggerations, omissions, repetitions, and incorrect dates: historic studies provide at best low and high figures for what actually happened during the great earthquakes of the past. Numbers of fatalities for even recent earthquakes can fluctuate wildly in the weeks following the disaster. Missing people may be included in the death toll when in fact, having lost everything, many simply have gone to stay with distant relatives. Others may be uncounted and buried among the ruins. Still others may succumb to the earthquake but only after a delay. As discussed in an earlier chapter, for many years the official death toll of the 1906 San Francisco earthquake numbered only in the hundreds, while a careful search of records revealed a total of close to 3,000 individuals whose lives were claimed—not necessarily right away—by the earthquake and subsequent fires.

The National Geophysical Data Center in Boulder, Colorado, maintains a list of earthquake fatalities that is reasonably accurate for the past 100 years but becomes increasingly conjectural in the preceding 3,000, providing uncritical references to many thousands of disasters. Some of the references to ancient earthquakes are culled from earlier compilations—from the great list of Robert Mallet and his son, for example, whose earthquake list of 1858 begins with three biblical earthquakes described in original Hebrew, Greek, and Latin manuscripts. Many lists of historic earthquakes repeat errors contained in earlier lists. These later compilations add false credibility to the earlier compilations, causing the unwary user to believe the repeated error provides corroboration for the first mistake.

In a few cases, catalogs persist in listing earthquakes that never occurred. The most notorious of these fake quakes is the Calcutta "earthquake" of 1737, in many popular books and Web listings still touted as the third most fatal earthquake in all history—with an inferred death toll of 300,000. A little investigation reveals that Calcutta in 1737 was visited by a flood that drowned, at most, 10 percent of its 30,000 population, and that no earthquake was reported by the surviving inhabitants. The exaggerated death toll appears to be an error in translation from a French account, whereas the report of an earthquake may have simply been the too literal acceptance of the metaphor describing the violent cyclonic wind buffeting Calcutta, which was responsible for blowing down the steeple of the church and part of a temple in this infant city.

Notwithstanding fake quakes and other limitations of historic earthquake catalogs, prior compilations do allow us to quantify the human losses from past earthquakes—and to project these calculations into the future.

Statistics of Fatalities

Not all countries have the same vulnerability to earthquakes because in many countries of the world earthquakes are really rather rare. However, those

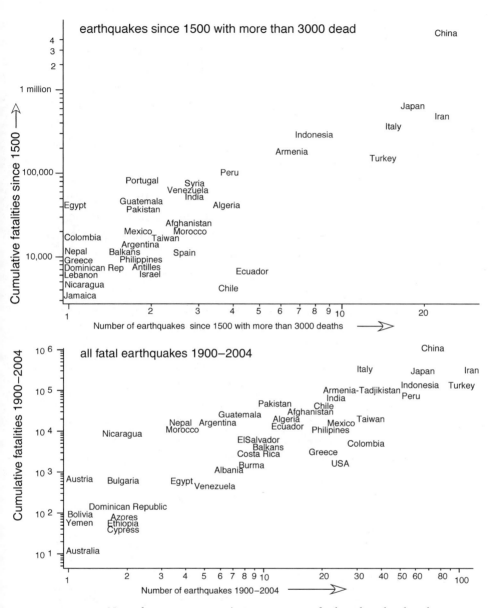

FIGURE 13.4. Not only are some countries more prone to fatal earthquakes, but the numbers of people killed by these earthquakes in these nations is much above average for both very large disasters viewed over several centuries, and for all earthquakes in the most recent century. Countries such as China, Japan, Iran, Italy, and Turkey are disastrous places to live. Most, but not all, earthquake disaster-prone regions on the right side of the two graphs are in the developing nations.

countries that have frequent violent earthquakes also have a history of huge numbers of fatalities from these earthquakes (figure 13.4). The distressing feature of these two graphs—one depicting fatalities from only large disasters in the past 500 years, the other showing all disasters in the most recent century—is that they show that these nations, in these two blocks of time, have not learned from their disastrous history. China, Japan, Italy, Turkey, Iran, and several other nations lost vast numbers of their people to earthquakes both before and after the development of earthquake-resistant structures in the 20th century.

If we are interested in possible trends in earthquake-related disasters, it is informative to plot the numbers of people killed by earthquakes as a function of time. Figure 13.5 reveals an astounding 8 million fatalities in the past 1,000 years, a number that can be shifted 10–20 percent up or down by selecting low or high figures for some of the larger historic earthquakes. These selections, however, do not alter the essential features of the graph.

The death march reveals a clear kink around the year 1500, the sudden apparent upward surge caused largely by the most disastrous earthquake in Chinese history (more than 500,000 people killed in 1524). Yet even with the event removed, there is a distinct increase in fatalities per year in succeeding centuries. The graph also shows relatively few earthquakes prior to A.D. 1000, presumably because only the most significant earthquakes were recorded, and only a subset of these records have survived. A further difficulty is that the histories of all nations are not evenly reported—compare the history of China with that of South America. Prior to 1600 the world had not even been fully explored. The change in the slope after 1500 thus reflects, at least in part,

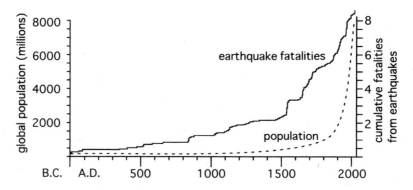

FIGURE 13.5. Earthquake fatalities vs. global population in the past two millennia.

an improved global history. But there may well be other reasons for the kink after 1500. It was at this time that printing was introduced in the West, for example, improving the chances of written records surviving. It could also be that the European voyages of discovery exported bad building practices (masonry structures) worldwide at this time!

In view of the myriad changes around 1500, we focus on data since this time, avoiding undue emphasis on the scant pre-1500 record. In figure 13.6 we look more closely at recent centuries to see whether earthquake fatalities increase in proportion to the number of people on the earth.

The resulting graph is surprising: a post-1800 factor of 10 increase in human population is clearly not accompanied by a tenfold increase in the number of people killed by earthquakes. The earthquake fatality curve is in fact almost linear—a straight line drawn through the jagged line would indicate 6 million fatalities in 500 years (15,000–25,000 earthquake fatalities

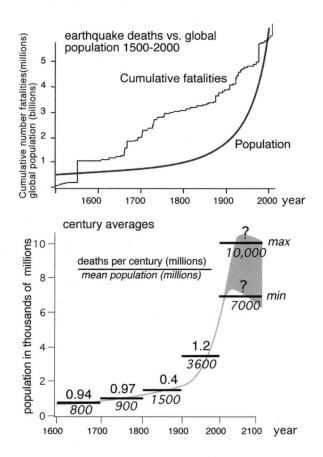

FIGURE 13.6. Earthquake-related deaths in the most recent 500 years show no simple relation to total global population, although without doubt the last century has been the most disastrous ever. Yet despite the increase in the number of earthquake-related deaths, the fraction of global populations killed by earthquakes has fallen. In the 17th and 18th centuries, earthquakes claimed 0.1 percent of the average world population; in the 19th and 20th centuries this fell by a third.

per year; see figure 13.7), with occasional upward jumps caused by deaths during individual catastrophic earthquakes that strike large centers of population. Taken at face value, the pre- and post-1800 curves suggest a paradox: that world population has been increasing without a commensurate increase in earthquake fatalities. Numerically, earthquakes killed 0.1 percent of the global population in the 17th and 18th centuries, but only 0.03 percent in the 19th and 20th centuries. Could everything we told you at the beginning of the chapter be wrong?

Could the world have effectively become a safer place, at least where earthquakes are concerned, over the last 200 years? This flies in the face of common sense: stringent earthquake codes are a late 20th-century invention at best, and even then in only limited areas. A resolution of the paradox much more likely lies in the ever-murky business of statistics, and the fact that the calculations discussed in this chapter depend critically on small data sets. Although the effects of large catastrophes are severe, in world history we have seen only 120 devastating events killing 25,000 or more, compared with some 5,000 earthquakes with smaller fatality counts. Perhaps 120 extreme earthquakes are statistically insufficient to reveal the true trends.

The statistics of the worst killer quakes—those claiming hundreds of thousands of lives—are clearly more limited still. The 2004 Sumatra earthquake alone has changed the fatality curves significantly. If an earthquake were to occur tomorrow and claim 800,000 lives, the curve would convey a very different impression tomorrow than it does today. Scientists have a phrase to describe this sort of situation: small number statistics. When results depend critically on a very few data points, conclusions can be tentative at best.

Still, it can be a tricky question to answer. Has the world gotten seismically safer when measured on a per-capita basis (i.e., the thesis of this chapter is wrong), or are the data simply too poor to reveal the underlying trends (our thesis is correct)? A measure of support for the latter possibility comes from a consideration of the curve since 1800: although small-scale fluctuations largely obscure any overall trend, the data do suggest an increasing rate of fatalities over the last two centuries.

Leaving the mercurial small-number statistics of the biggest killer quakes aside for the moment, we next inspect only the 2,000 less severe earthquakes in the past five centuries—those earthquakes that killed fewer than 25,000 people and more than 10 people. (If this seems like a curious thing to do, bear with us.) In this graph we exclude, for example, the Bam earthquake of 2003 that resulted in death toll of 30,000, but we include the Bhuj 2001 earthquake

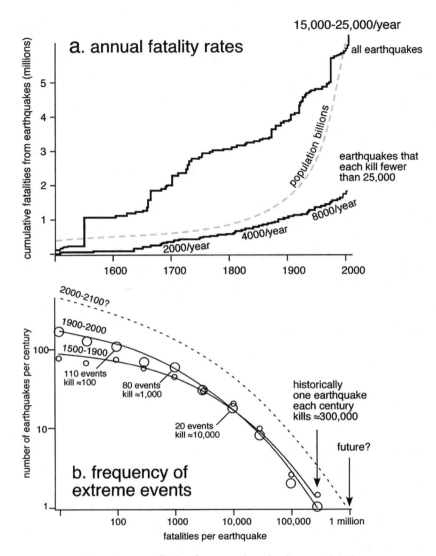

FIGURE 13.7. Attempts at predicting future earthquake disasters. (a) The uneven cumulative fatality curve for all earthquakes in the past 500 years predicts 15,000–25,000 fatalities per year from future earthquakes, although the prediction is meaningless for periods of time less than several centuries. In contrast, if the rare larger disasters are removed, we obtain a reasonably reliable prediction of 8,000 per year from earthquakes that each killed fewer than 25,000 people. (b) The numbers of earthquakes per century that kill specified numbers of people provides another way to glimpse our earthquake future. Based on the past few centuries, we might expect one earthquake to kill 300,000 people each hundred years. Global population increase (dashed line) suggests that a future earthquake could kill more than a million people—an event unprecedented in human history.

and the Izmit 1999 earthquakes that each resulted in 16,000–18,500 fatalities. While this cutoff is somewhat arbitrary, the resulting graph (the lowest line in figure 13.7a) is interesting because, unlike the erratic preceding graphs, the fatality count is well behaved, and well-behaved graphs can with reasonable reliability be extrapolated into the future. When this graph was first made (in 1995), it revealed that 5,860 ±500 people have died each year as a result of earthquakes, and that this rate was increasing as a result of global population increase. Accounting for a global increase in populations, the calculations suggested that about 8,000 people might die from earthquakes in the next few decades. Since this prediction was made, the number of people who have been killed in this type of earthquake disaster—"modest catastrophes"—has exceeded 10,500 people/year. The higher numbers may not necessarily mean a worsening in the human predicament, because fewer than ten years have elapsed since the prediction was made, and fluctuations in rate are expected on short timescales.

The reader might still be puzzling over the idea of separating the modestly catastrophic earthquakes from the truly devastating events. Such manipulation of statistics by excluding the most disastrous events is in fact rather arbitrary, but it has the benefit of guiding our thinking about what might be called a background rate of earthquake fatalities. That is, by excluding the rare and unpredictable catastrophic events, these calculations can tell us, apparently with a fair measure of certainty, how many earthquake fatalities are likely in a given year. In a sense, this number represents a good lower bound on the number of deaths expected per year: the number in any given year is unlikely to be much lower, while it will be much higher in those years when catastrophic events do strike.

So what, then, of the largest events, like the Tangshan earthquake of 1976 or Sumatra in 2004, each with more than 250,000 fatalities? Can we make any forecasts of the likelihood of just one of the world's supercities being hit in the next century? The overall global rate of megaquakes can be estimated, but the timing of the largest earthquakes cannot be predicted precisely. It is even less possible to predict when one of these megaquakes might score a direct hit on a major population center. We can, however, explore the following question: Given statistics of past earthquakes, how many people could be killed in a large future urban earthquake?

A rough answer to this question can be obtained by plotting the number of earthquakes that kill a specific number of people in a given length of time (figure 13.7b). We choose a century as the time interval; ask how many earth-quakes occurred that caused 100 fatalities, how many caused 1,000, and so on; and plot a graph of these numbers to see whether any obvious relation-ship emerges. (Scientists construct such plots routinely nowadays, because very often an obvious relationship does emerge—one that reflects the nat-ural mathematical hierarchies of many natural systems.)

In figure 13.7b, the smoothly down-dipping curves link the greater num-bers of low-fatality earthquakes with the smaller number of very lethal earthquakes. Curves are not as useful as straight lines if we are to base fore-casts upon them, but the curve is reasonably well-behaved in that the aver-age curves (and rates) for the period 1500–1900 more or less mimic the 20th-century data. The curve for the earlier period falls below the 20th-century curve at the low end, as expected, given the near certainty that we do not have complete information for smaller earthquakes from the pre-19th-century times. But does the resulting graph hold any clues for our future? The curves based on past data reveal that in the past the planet has experienced an earth-quake roughly once a century that killed 300,000 people, suggesting that similar events will strike in future centuries. Or do they?

Recall that the data on which this graph is based come from a world pop-ulation that averaged roughly 3.6 billion in the 20th century and less than half this in preceding centuries (figure 13.1). During the 21st century, whether we take the United Nations' low or a high estimate, the earth will be home to two to three times as many people—potential earthquake targets, all.

If we make the simplest correction for the expected population increase, we effectively raise the curve (dashed line in figure 13.7b). This suggests that three or more 300,000-fatality events may be possible in the 21st century, and regretfully one of these has already occurred. But the curve suggests some-thing new: that 600,000 to 1 million people could be killed in a single earth-quake (indicated by the point where the upper dashed line intersects the once per century axis—the bottom line).

It is perhaps rash to predict something that has no precedent in human history. Never has an earthquake disaster caused a million fatalities, but then again, never before have we presented earthquakes with such a large human target. Not only have we increased our population, we have concentrated people into cities of unprecedented size, and placed many of them in loca-tions certain to be shaken by future large earthquakes. Given that the Tang-shan earthquake resulted in the loss of 20–30 percent of the total population,

a direct hit on one of our current 27 megacities could very plausibly generate a mega-fatality count. A 1-million fatality count is less than 10 percent of
the population of many of our largest earthquake-prone agglomerations.

However, the above calculation is a very simple one, especially if we return to the surprising result from the statistics of figure 13.5: if we take a two-
century average, the death toll from earthquakes in 1600 to 1800 exceeds the
death toll from 1800 to 2000. Still, while one can debate the statistical issues,
it remains almost self-evident that future catastrophes will be increasingly
severe when large earthquakes strike increasingly densely populated urban
centers—especially the large number whose structures remain vulnerable
or, worse yet, are becoming increasingly vulnerable.

To remain in the murky realm of statistics a bit longer, we consider some
data from the 20th century. If one plots earthquake fatalities versus magnitude for quakes between 1900 and 2004, one finds, as expected, a considerable degree of variability (figure 13.8). An M7 earthquake in a remote part of
Alaska can kill zero people; the same magnitude in India (or perhaps New
York City) can kill tens of thousands. Nevertheless, the upper edge of the
cloud of points essentially defines the worst-case scenarios: temblors of a certain magnitude that have scored direct hits on densely populated and highly
vulnerable cities. Defining a precise mathematical line to this upper edge
would be difficult. However, the data suggest that an earthquake of M7.5 or
greater might someday unleash a fury sufficient to kill a million people. That
is, the data suggest that the ultimate killer quake will not necessarily rank
among the planet's very largest earthquakes. In fact, since M7.5 events occur
far more frequently than M8.5 quakes, it may actually be likely that the ultimate killer quake will be a relatively smaller earthquake.

Once we admit that unprecedented death tolls are possible, speculation
naturally follows: Which of our supercities or megacities will next be hit by
a large earthquake? This cannot be answered by the kind of statistical approach we have developed in this chapter. In fact, it really can't be answered
at all, given our current inability to predict future earthquakes. But a short
list is of possible targets can easily be constructed from the past 1,000 years
of earthquake data and U.N. predictions concerning the growth of our world
cities (figure 13.9).

Some cities are literally braced, ready to be hit by a large earthquake. To
name a few: Tokyo, Wellington, Los Angeles, San Francisco. Many others are
partly prepared—while admitting that a large earthquake is likely, other priorities dominate their agenda. Most of these cities are in the developing na-

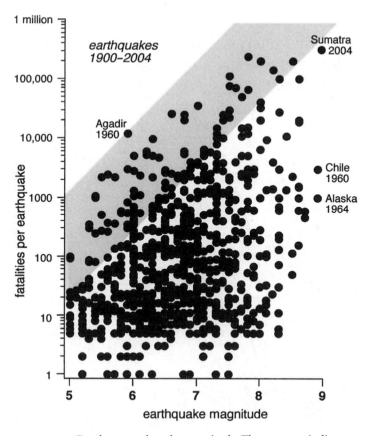

FIGURE 13.8. Deaths vs. earthquake magnitude. The gray area indicates upper and lower estimates for the number killed by a given size earthquake. Thus, for an M6 earthquake more than 10,000 people can be killed if the earthquake is a direct hit on a poorly built city (e.g., Agadir, Morocco, 1960), though usually fewer than 100 are killed, and in many earthquakes, none are killed (not shown on the graph). A million-fatality earthquake would most likely be caused by an M7 to M8 earthquake close to a megacity.

tions, and while they may have enacted building codes, their implementation is by no means consistent: Tehran, Caracas, Istanbul, Delhi. In these countries, earthquake engineers are actively engaged in efforts to persuade leaders to enforce codes, but the notion of seismic risk is often balanced against other pressing national needs.

The large cities in Figure 13.9 are ranked according to their estimated sizes in 2010. These are cities that have been damaged by a historic earthquake in

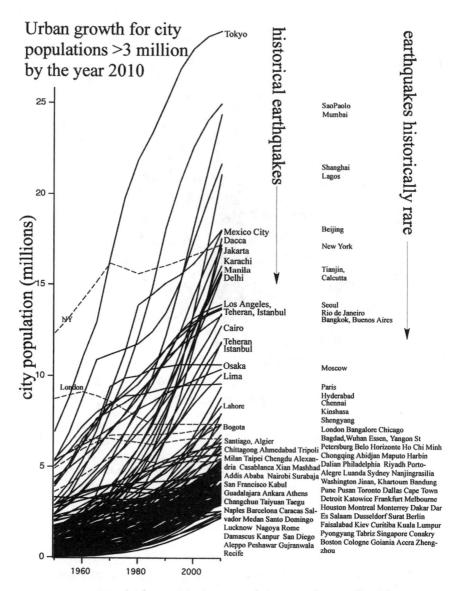

Urban growth for city populations >3 million by the year 2010

Tokyo

historical earthquakes

earthquakes historically rare

city population (millions)

25 —

20 —

15 —

10 —

5 —

0 —

NY

London

1960   1980   2000

Mexico City
Dacca
Jakarta
Karachi
Manila
Delhi

Los Angeles,
Teheran, Istanbul

Cairo

Teheran
Istanbul

Osaka
Lima

Lahore

Bogota

Santiago, Algier
Chittagong Ahmedabad Tripoli
Milan Taipei Chengdu Alexan-
dria Casablanca Xian Mashhad
Addis Ababa Nairobi Surabaja
San Francisco Kabul
Guadalajara Ankara Athens
Changchun Taiyuan Taegu
Naples Barcelona Caracas Sal-
vador Medan Santo Domingo
Lucknow Nagoya Rome
Damascus Kanpur San Diego
Aleppo Peshawar Gujranwala
Recife

SaoPaolo
Mumbai

Shanghai
Lagos

Beijing

New York

Tianjin,
Calcutta

Seoul
Rio de Janeiro
Bangkok, Buenos Aires

Moscow

Paris
Hyderabad
Chennai
Kinshasa
Shengyang
London Bangalore Chicago
Bagdad,Wuhan Essen, Yangon St
Petersburg Belo Horizonte Ho Chi Minh
Chongqing Abidjan Maputo Harbin
Dalian Philadelphia Riyadh Porto-
Alegre Luanda Sydney Nanjingrasilia
Washington Jinan, Khartoum Bandung
Pune Pusan Toronto Dallas Cape Town
Detroit Katowice Frankfurt Melbourne
Houston Montreal Monterrey Dakar Dar
Es Salaam Dusseldorf Surat Berlin
Faisalabad Kiev Curitiba Kuala Lumpur
Pyongyang Tabriz Singapore Conakry
Boston Cologne Goiania Accra Zheng-
zhou

FIGURE 13.9. Growth of supercities (2010 population exceeding 3 million) from U.N. predictions. The dashed lines show slow growth in developed nations. Approximately 40 percent of these 120 cities are in earthquake-prone regions. More than 70 percent of these vulnerable cities are in the developing nations where earthquake-resistant construction is often a low priority.

their past, typically because they are located on or near an active plate boundary. We omit the much larger number of cities potentially at risk from earthquakes that are far from plate boundaries because we do not have a good earthquake history in their location. Such cities would include such perhaps unlikely names as New York, Boston, Denver, and Chicago, to a name a few in the United States. The most striking observation is that more than half of the most vulnerable cities are in the developing nations. This is of concern, because these are the regions where expensive retrofitting projects are least tenable.

Returning to 20th-century data, if one plots the number of earthquakes in different countries versus the total number of fatalities in each country, the results highlight the uneven distribution of earthquake vulnerability (figure 13.4). Earthquakes have been especially costly, in terms of human lives, in China, Iran, Turkey, Armenia, Indonesia, and India. Among industrialized nations, only Italy and Japan have experienced heavy 20th-century losses.

The United States falls notably below the norm for countries with similar earthquake rates. The fatality count would be higher if one used the high figure (~3,000) for San Francisco in 1906, but still the overall number would be low. To some extent this reflects the success of risk mitigation efforts in seismically active California. Temblors such as those in southern California in 1971 and 1994 claimed remarkably few lives; clearly building codes had a lot to do with this. But to some extent the cumulative total may reflect sheer dumb luck. A lot of U.S. earthquakes occur in a state (Alaska) with very few people. Moreover, the fatality count may well reflect quite a bit of happenstance: the fact that big central and eastern U.S. quakes happened to strike in the 19th century rather than the 20th, the fact that the last great earthquake in California was in 1906.

Modern loss estimation tools have been developed to predict both dollar and human life losses from future earthquakes. Such calculations take into account not only predictions of ground shaking but also building inventories and vulnerabilities. For any number of plausible earthquake scenarios in California, the predicted death tolls run easily into the thousands—even the tens of thousands. The Pacific Northwest represents an even greater potential time bomb: When the region is hit by the next great earthquake like the one that struck in 1700, it will be rocked by effects similar to those from the 2004 Sumatra quake. Still, while the United States may eventually experience more losses than the graph reflects, most of the countries at the top of the curve are very likely to remain in their unfortunate relative positions.

A catastrophic megaquake is inevitable only if we continue to construct cities that collapse in earthquakes. However, it is largely in the developing nations that the largest building program in the history of mankind is now under way. More dwelling units are under construction in the world at present than at any time in our history. Even the massive construction programs after the two world wars did not plan for the numbers of new human beings who will be arriving in coming decades. Our planet is about to double its population for possibly the last time—from 6 billion to possibly 12 billion people. This will require approximately 1.5 billion new dwelling units. It is surely appropriate to include in these new structures a modest degree of earthquake resistance. In the next chapter we show that an investment of perhaps 10–20 percent of the initial construction cost of a building is all that is required to make it relatively safe from the violent accelerations of nearby earthquakes, a much smaller investment than retrofitting a billion and a half unsafe structures.

As noted, curves such as those shown in figure 13.7 are common in science: in a given area, numbers of earthquakes follow a similar curve if plotted against magnitude. That is, if there are, on average, 100 M4 earthquakes in a given region in a year, one expects about 10 M5 and 1 M6 earthquakes to occur as well. A precise forecast of future fatalities is open to debate, but it defies all common sense to imagine that future death tolls will not be larger—possibly much larger—than those that have already happened. If a 20th-century earthquake in China killed hundreds of thousands of people, one can scarcely imagine that future earthquakes won't someday be worse. To change this grim forecast in any meaningful way, we must change the equation. In our final chapter we discuss the unique opportunity that is now before us to do just this.

14

The Age of Construction

Not to quarrel with the intelligence that reads God behind
seismic disturbance, one must still note that the actual damage
done by God to the city was small besides the possibilities for
damage that reside in man-contrivances; for most man-made
things do inherently carry the elements of their own
destruction.

> — Mary Austin, "The Temblor," in *The California
> Earthquake of 1906*, edited by David Starr Jordan
> (San Francisco: A. M. Robertson, 1907), 358

Earthquakes don't kill people — buildings do.
> — Nick Ambraseys

As the human population of our planet rises to hitherto unprecedented levels, we find ourselves wondering whether the half-century from 1990 to 2040
might be remembered not so much as the age when the oil ran out, as *the age
of construction*. Never before have we built so many dwellings, roads, dams,
and civic structures than will be constructed during the span of this half-
century. A little reflection suggests that in our (allegedly) highly evolved society, with our sophisticated knowledge of the forces of nature and the
strengths of materials, we would be stupid to commit the unforgivable sin of

knowingly constructing buildings that will crush and maim our descendants. Yet in many parts of the world this is indeed what we are doing.

Homo sapiens decided long ago to live in houses. Other animals do it, but rarely do they build such precarious structures as do humans. The nests of birds are woven to be resilient, mammals and reptiles live in caves selected for their permanence, burrows are dug by animals content with the knowledge that a little more burrowing is all that's needed to keep the walls in place, or the driveway clear. Only humans spend at least eight hours of every turn of the planet within a dwelling assembled from a variety of materials that are often close to the point of structural failure, and often without considering the consequences of constructing permanent dwellings in regions subject to geologically extreme events. The shift from *Homo* the hunter-gatherer to *Homo urbanensis* means that many of the remaining 16 hours of each day are spent in another structure, more often than not also assembled with an eye on thrift —maximum volume for minimum cost. Even the journey to and from these different structures can expose humans to seismic risks—as is evident from the collapse of bridges and overpasses in recent earthquakes.

The damage done by an earthquake is caused by shaking, either directly or indirectly (via landslides, etc.). Shaking involves accelerations: the rate at which speed changes or, in qualitative terms, what can be thought of as "jerkiness." Close to the start of the human population explosion, Isaac Newton pointed out that accelerations acting on mass result in forces. These forces, which are directly proportional to the mass of the building, can destroy a building. In a very literal sense, then, buildings have the ability to destroy themselves.

Structures are always designed to resist the forces of gravity (the underlying reason why walls are vertical). An inclined wall will eventually topple under its own weight. A plumb bob points to the center of the earth, and the walls of all our buildings follow the line of the string. Most builders have infinite faith in the assumption that the plumb bob will always point to the center of the earth. They would be surprised to learn that during a sufficiently large earthquake, the plumb bob will whip horizontally to several points of the compass, and sometimes briefly even to the heavens above!

Although it is perhaps an absurd thought, many buildings would survive earthquakes if their builders imagined they were constructing their dwellings on the side of a vertical cliff, or suspended from the roof of a cave. The accelerations on such hypothetical dwellings, resulting from the simple tug of gravity, resemble in some respects the forces encountered in an earthquake.

The acceleration we call gravity is so built into our psyche that its friendly pull is easily ignored. Most builders have an intuitive feel for how thick a beam is needed to hold up to a roof to resist the pull of gravity. Where intuition fails—for example, when a heavy snowfall causes collapse—rebuilding brings with it stronger roof beams and a sturdier reconstruction.

With earthquakes, however, the builder usually gets only one test, and many never even have to take this final exam. Those who do pass or fail are unlikely to learn from the experience, or to be punished for their failure or praised for their success. There is no official examination committee for buildings that experience earthquakes, although after each major earthquake the world's structural engineers and seismologists take a good look at what structures performed well and which did not. The lessons they learn are generally the same. Many structures collapse for very simple reasons, and although some building collapse is the result of local geologic conditions, many failures are common to the new structures in cities the world over.

Large, destructive earthquakes are sufficiently infrequent that many—even most—generations will not experience one, even in active plate boundary regions. When such an event does strike, its lessons will be lost if that generation is unable to tell others of a fatal design flaw. Thus mankind has persisted over the ages in erecting highly vulnerable structures. In the Information Age, we have the opportunity to do better. In this chapter we look at a world of buildings with the nervous eye of the earthquake engineer. We start with some lightweight structures, then mention common failings of more massive structures made from adobe, bricks, and concrete.

Grass Houses versus Stone Houses

For many societies the most abundant building materials for dwellings are bamboo and grass, wattle-and-daub, straw, twigs, cloth, and poles. These structures have little mass compared with their strength, and they survive even the strongest shaking with impunity. Their rare collapse will likely cause the inhabitants the trauma of a yawn—time to fix the roof, to patch the wall, to repair the lintel.

Lightweight construction is practiced in many societies by necessity rather than by intent. Flimsy but tough fronds and twigs may be all that's available in some societies or, more often than not, the only materials that people can afford. Figure 14.1 shows the style of building adopted in two corners of

FIGURE 14.1. Traditional construction practices in India have not changed much from 1860 to the present. The two top images are from 1860; the two bottom images are from 2004. The bottom left image shows a Kangra roof in the Himalaya and the bottom right image shows a dwelling on the Shillong Plateau. (Top two images from Caleb Wright and J. A. Brainerd, *Historic Incidents and Life in India* [St. Louis, Mo., 1861]; bottom two images, Roger G. Bilham)

India—the Kachchh Peninsula and the Khasi Hills on the northern edge of the Shillong Plateau—both regions that have repeatedly experienced mighty earthquakes. We choose these two regions to illustrate climate extremes: the Khasi Hills, site of the 1897 Assam M8.0 earthquake, with almost 10 meters of rain per year, is one of the wettest places on earth, whereas the Kachchh region, site of the 2001 Bhuj M7.6 earthquake, is one of the driest. They both experience warm winters and very hot summers. In both regions rocks are not readily available. In both regions the wealthier segment of the population has chosen to use their means to import stones and to construct (often lethal) masonry structures—or, more recently, to construct dwellings of concrete and steel.

Lightweight construction methods are used in the slums surrounding large urban centers in countries such as India. These shantytowns in the developing world are assembled from a sad mix of corrugated iron, hammered oil drums, and plastic sheets. The single-story dwellings keep out the rain and the sun and are home to many millions (figure 14.2). They are intrinsically safe from shaking because when they fall their collapse is seldom fatal to their occupants.

However, sprawling slums grow on the sides of cities in areas where the wealthy have decided it is either unsafe or undesirable to live—in swamps and on unstable hillsides. Here, earthquake dangers are different. Swampy ground may lose its internal cohesion and liquefy or flood during earthquake shaking or in its aftermath, and hillsides may collapse. The ever-present risk of disease can only be exacerbated when earthquakes disrupt what little infrastructure stability and order existed in the first place.

The epitome of lightweight construction is the use of timber frameworks clothed with plywood sheeting, one of the most common construction methods in the urban agglomerations of the United States. Wooden structures have built-in resilience. The two-by-four skeleton of timber is typically clothed with plywood, whose layers of lightweight fibers resist extreme lateral forces. Wood frame structures are tough (as are the trees from which they are derived); they bend and creak, but they rarely fail catastrophically during an earthquake. (Trees have been known to break in two during earthquakes, but only when subjected to extremely severe shaking.) The worst that normally occurs is that objects within wood-frame buildings will be thrown around, and the structure may depart from its foundations. The secondary effects of a tossed but intact building are that survivors (shaken and bruised) often must contend with fires, gas leaks, and possible explosions accompanying the rupture of underground pipes. Water pressures strong enough to quell these fires may not be available because of the fracture of underground pipes, making the danger from fire worse than the danger from building collapse. The fatal flaw of timber is that it is inflammable. In California, where earthquakes are an acknowledged real and present danger, many homeowners have invested a few hundred dollars for a simple seismometer device that will shut off a home's gas valve if strong shaking occurs.

FIGURE 14.2. Structures like this mud-and-lathe dwelling survived an M5.1 earth-quake in Brazil in 1986, whereas nearby concrete skeleton buildings were ruined. Government engineers incorporated this shake-resistant design in the reconstruction of dwellings after the earthquake. (Roger G. Bilham)

Masonry: Sticks and Stones

Where wood is too expensive or simply unavailable—in deserts or in mountainous regions above the tree line—the options for lightweight construction dwindle. By default, in such locations humans have assembled dwellings from stone, or from man-made stones: bricks made from clay (or ice). The ability of a pile of bricks or stones to resist violent shaking depends on the adhesion between the bricks as well as their internal toughness. A building's survival can also hinge on the direction of shaking, and its duration.

In addition to the brick buildings found in many cities, masonry includes the fine cut-stone blocks of cathedrals and courthouses. At the least desirable end of the scale, the weakest masonry construction consists of slender walls assembled from round river boulders cemented with mud. Structural integrity in an earthquake is increased when the boulders are angular, with few or no rounded surfaces and where the mud is replaced with strong cement. Earthquake survival becomes increasingly likely with dressed stone blocks, especially if these are massive or specially fit to their neighbors (Inca style). The Roman foundations of Baalbeck include three finely cut rectangular blocks, each measuring 4.6 by 4 by 20 meters and weighing 800 tons. Recognizing the futility of adhesive between heavy blocks, Greek and Roman architects sometimes used thin layers of lead instead of cement between sections of tall columns. These lead washers cushioned the highly stressed points of contact between blocks, and absorbed some of the stresses during earthquake shaking.

But few masonry buildings use the massive cut stones of classical architecture. Most are assemblages of bricks that can be picked up in one hand and simultaneously cemented into place with a trowel wielded in the other. Without special reinforcement, masonry structures fail so predictably in earthquakes that the degree of damage is a key factor used by seismologists to study the severity of earthquakes. One of the indicators of the severity of shaking between intensity VI and intensity VIII of the famous Mercalli intensity scale is the number of masonry chimneys that collapse. Weak chimneys start to crumble during intensity VI accelerations, and the toughest masonry chimney is usually gone by intensity VIII. These accelerations vary from one tenth to one third the acceleration due to gravity (figure 14.3). Chimneys are such good indicators of shaking severity because they are simple structures: slender, heavy objects whose height far exceeds their width.

FIGURE 14.3. The M6.7 Northridge, California, earthquake of 1994 damaged the chimney of this typical wood-frame house, in the town of Reseda, which otherwise escaped with no structural damage. (Susan E. Hough)

FIGURE 14.4. The M6.3 Santa Barbara earthquake of 1925 caused substantial damage to unreinforced masonry buildings. The hotel shown above lost large portions of its outer walls, which fell away from the inner structure, leaving rooms and furniture exposed.

A brick wall is narrow in one direction but long and squat in the other; in an earthquake a wall can fall only in one of two directions. A chimney can fall in any direction, caring little whether it collapses into or away from a house. The bricks of a masonry chimney are seldom cemented together strongly enough to resist their separation when earthquake shaking reaches a certain threshold. Taking account of both chimneys' vulnerability and their uniformity of design, chimney damage has in some cases been used as a direct proxy for shaking severity within an urban area, for example, in and around Seattle following the 2001 M6.8 Nisqually earthquake.

Withstanding separation is the key to earthquake resistance (figure 14.4). The essential feature of a masonry structure is that it has high strength in compression, but is weak in tension—that is, such a structure can support a lot of dead weight, but its structural elements are easily separated by active forces. Thus, masonry structures can be strengthened by steel reinforcement, either internal (figure 14.5) or external, with steel bands wrapped around masonry elements. Simply put, the elements of masonry structures must be prevented from drifting apart. Many medieval structures are currently held

FIGURE 14.5. (Left) A postcard illustrates construction of a "fire-proof and earth-quake-proof" building in the aftermath of the 1906 San Francisco temblor. (Right) A modern retrofitting of the U.S. Geological Survey building in Menlo Park, California. (Susan E. Hough)

together by steel bars, bands, and bolts to prevent them from collapsing from old age. In the world of earthquake engineering, the retrofit of masonry structures to improve their resistance to shaking often takes the form of steel bands wrapped around the top of the house, and sometimes its waist. Although additional diagonal bracing is the most effective retrofit measure because such bracing will prevent the potentially devastating shift of the top of a building sideways relative to its base, this approach is expensive and seldom considered architecturally acceptable (figure 14.6). Some architects have bitten the bullet and included X bracing as an architectural feature in front of, or behind, the traditional rectangular windows through which we view our world.

The incorporation of belts and bands to hold structures together is not a modern notion. Classical architects in Greece and Rome and Gujarat held their structures together with iron bands and pegs. In remote villages far from the influence of architects and where limited supplies of wood may be available, tie beams have been incorporated into dwellings. A possible reason for this interesting adaptation is that these are the fittest buildings to have survived previous earthquakes. Dwellings in the Himalaya often include a layer of timber between the stories (figure 14.7). The wood acts as a continuous band that by friction alone prevents the stonework above and below from shifting in an earthquake, thereby imparting to the structure the ability to wobble rather than collapse. Many buildings that survive earthquakes unscathed in Kashmir and Katmandu have done so because they have included balconies supported by through-going beams that (unwittingly) held the masonry walls together during shaking. Regrettably, many newer buildings omit or skimp on these traditional through-going beams, reserving wood for internal floor beams and pillars, thereby inviting disaster in future earthquake events.

The inherent elasticity of wood-frame/masonry infill buildings in Turkey (figure 14.8) enables them to survive earthquakes that destroy modern concrete structures. The masonry fill, bonded by weak cement, grinds and crumbles during shaking, absorbing some of the stresses imparted by the quake and requiring post-earthquake cosmetic repairs, but few funerals.

Thus the old adage "Sticks and stones may break my bones . . ." is reversed in earthquake country, where the incorporation of minor amounts of sticks can often prevent the collapse of stones. As the availability of wood has been reduced, however, new structures have been unable to emulate these traditional solutions. In the 1993 Latur earthquake, which killed 7,500 people in south-central India, it was clear that older buildings survived better because they used more timber in the walls and roof than buildings constructed more recently. These later structures also tended to be constructed more hastily, using rounded river boulders for the walls instead of angular partly dressed stone.

Adobe block structures are the weakest type of masonry construction, but even for these there are remedies to mitigate the effects of shaking, espe-

FIGURE 14.6. The brick building on the left in Shillong, India, is being assembled with bricks placed on their sides (at the new owner's request to save money). In contrast to this unstable pile of bricks, the owner's present traditional dwelling (shown on the right) will survive the next earthquake in the region. (Roger G. Bilham)

FIGURE 14.7. Traditional buildings constructed where wood is scarce and stone is abundant. (Left) A structure in the Himalaya uses horizontal planks (shown by arrows) to bind courses of rock together. (Center and right) Newly dressed masonry structure in Nepal assembled without mortar. The only throughgoing plank holding the masonry together is a thin strip (designated by an arrow) that is unlikely to hold up in an earthquake. (Left photograph, Nick Ambraseys; other photographs, Roger G. Bilham)

cially if the objective is to provide those few vital seconds needed for a family to escape into the open. The seismic retrofit of adobe buildings ranges from drilling thin holes into each clay block and tying it to its neighbors with plastic bags, shoelaces, or nylon string (for owners with plenty of time at their disposal), to inserting tension rods through the entire wall structure (for historically important buildings). Numerous ways have been devised to toughen new adobe bricks, using chemical or fiber additives, but these are rarely available to the world's poor, for whom construction (or reconstruction) with advanced materials is not an option.

The replacement of an adobe mud roof with one made from insulating plastic, corrugated iron, or wood, an apparently obvious way to reduce one's risk of being buried alive in a future earthquake, is usually unaffordable to most adobe dwellers. Even when a government provides materials and guidance, there is usually strong resistance to reducing the thickness of an adobe mud roof, which is considered not just insulation but also a thermal reservoir that buffers extremes of heat and cold. It is usually only after an earthquake that the serious consequences of living beneath many tons of mud become apparent. Sadly, the lesson is soon forgotten. Although a family may live in tents for a few years, adobe structures begin to grow, first for storage, then for livestock, and finally for new families. After a generation and a half, a village more often than not returns to its pre-earthquake habits.

Concrete Skeletons: High-Tech Structures in a Low-Tech World

Wood has become increasingly scarce in the developing world, and given that bricks require wood or coal to be fired, these, too, are increasingly unavailable. For this reason most multifamily structures are now made from concrete cast around a spindly assembly of steel rods. The body of this concrete skeleton is fleshed in with a skin of masonry, with the occasional window opening and door. The walls may not be real bricks but, instead, hollow ceramic tile or cinder blocks. The assembly of such structures has an elegant

FIGURE 14.8. In some parts of the world, traditional architecture utilizes timbers with masonry/rubble infill. This photograph was taken days after the 1999 Duzce earthquake in Turkey. The traditional building (background, left) survived, while its more modern neighbor (background, right) did not. (Roger G. Bilham)

simplicity, but there are numerous places in a low-tech world where this high-tech system can go wrong.

The quality and quantity of materials needed to erect competent concrete frame structures are very different from those required to guarantee the continued competence of a building during the jolts and capricious shaking encountered in an earthquake. Rarely does the average contractor have an intuitive feeling for the unusual loading that may occur three seconds or three minutes into an earthquake. The concrete skeleton must resist being deformed rapidly, its tidy rectangular elements twisted in an agony of parallelograms created by sudden vertical and horizontal forces. These forces concentrate at the joints between floor and ceiling, where the unwary contractor may have concentrated his assembly skills the least (figure 14.9).

Construction of a concrete frame dwelling starts with a foundation from which protrudes a forest of reinforcing bars that will support the next floor. If an engineer is in attendance, the thickness and quantity of each bar are prescribed by calculations of building height and anticipated loads. But if no engineer is present, the temptation to use less expensive steel (thinner and/or more brittle), or to use a thinner mix of cement to encase the steel, can be an irresistible attraction to increase profits.

The correct blend of steel and concrete is a marriage made in heaven. Concrete has great strength in compression, but none in tension, whereas spindly steel bars have great strength in tension but none in compression. Yet all too often a fatal combination of brittle steel and low-strength concrete is used. Brittle steel snaps in tension, and weak concrete ruptures in compression— a marriage made in hell.

A common misunderstanding among contractors is that the little metal collars (stirrups) wrapped around the vertical bars inside a concrete column are there simply to hold them in place while the cement cures. Few realize that these are not simply spacers, and that with insufficient thick stirrups a concrete column is doomed to explode, crush, and collapse in an earthquake. In new construction in the developed nations, a long spiral spring of steel is used to encircle the vertical bars, instead of the labor-intensive hand-tied stirrups.

Another point of failure is the presence of air bubbles in the cement, or the addition of dirt, instead of sand, to the mix. Holes in the cement are especially dangerous if they occur at corner joints between floor and ceiling, where shaking concentrates the forces of destruction. Most house buyers or

FIGURE 14.9. Incorrect assembly of concrete and steel skeleton buildings can result in catastrophic failure either on the ground floor (two images on left) or from floors merging into each other, or "pancaking" (two images on right). Poor-quality concrete, insufficient steel reinforcement, and flimsy stirrups were the causes of these buildings' collapses. All damage was caused by the 1999 Duzce earthquake in Turkey. (Roger G. Bilham)

renters assume that concrete frame buildings have been assembled correctly. Even were they suspicious of their dwelling, they would have no way to examine its invisible internal structure. There is no hospital where invalid buildings can be taken for a checkup. Aluminum cans and builders' rubble were found in the reinforced concrete walls of one of the dwellings that collapsed in the 1999 Izmit, Turkey, earthquake.

The final insult to the integrity of a concrete frame structure is the incorporation of large windows, especially at street level. The lowest floor of a multistory building is the one subjected to the greatest stress. All too often, though, it has the least strength because a shop front or garage door weakens the ground-floor wall. This "soft" lower floor is the undoing of many

FIGURE 14.10. (Left and center) These concrete frame structures in Kathmandu reach for the sky yearly as owners circumvent height restrictions. (Right) A concrete structure in Caracas has a weak third story that would collapse in an earthquake, causing the overlaying nine stories and also the bottom two stories to pancake into each other. (Roger G. Bilham)

otherwise safe concrete frame structures (figure 14.10). During an earthquake the ground floor collapses and the building is brought to its knees, or it pancakes like a stack of cards.

Foundation Failure, Landslides, and Tsunami

The litany of construction errors is not limited to the area above ground. Some parts of the world are simply not a good place even to start construction—no matter how much integrity a structure might have. Areas prone to landslides can usually be identified from local conditions or from a history of previous disasters. Areas of land near rivers or close to sea level may also be inadvisable because of their ability to liquefy during an earthquake. Heavy buildings will sink and light structures, like gasoline tanks, will rise from the ground during persistent shaking. Underground utilities—water, gas, and electricity lines—rupture during shaking. Unfortunately, many of our cities have been constructed in such environments, and the design of safe structures in such conditions requires carefully engineered and costly foundations (figure 14.11). Many Pacific ports are located where tsunami damage may occur from nearby or distant earthquakes. Many Japanese villages have

FIGURE 14.11. These buildings demonstrate insufficient attention to foundations. (Left) Earthquake shaking liquefied the foundations supporting these buildings in Niigata. (Right) An earthquake could spill this otherwise sturdy building down its hillside in the Khumbu, Nepal. (Left: National Geophysical Data Center; right: Roger G. Bilham)

accepted the danger from tsunami and have constructed special dams facing the sea to prevent damage from large waves.

Room for Optimism?

Our gloomy view of dwellings is colored by visits to areas of recent earthquakes where apparently identical buildings collapse or survive depending on the quality of their assembly. We prefaced this chapter with Nick Ambraseys' observation that buildings kill people, not earthquakes. Yet it is clear that the real culprit is often not the building so much as the builder. The knowledge of how to assemble a structure so that it will not fall down has been around for decades, but all too often houses are assembled by contractors who are neither residents nor engineers; they are in business to make profits in a society that, especially in the developing nations, is driven by the low bid. This, of course, is a recipe for disaster. Back in 1907, Charles Derleth, Jr., observed, "Any one who has carefully studied earthquake destruction can not fail to appreciate that great structural losses are due primarily, except in the immediate region of a fault line or upon loose deposits, to faulty design, poor workmanship, and bad materials; let us hope through ignorance and a blind disregard for earthquake possibilities; yet I regret to add that I fell con-

vinced that much of the bad work is due to a combination of criminal care-
lessness, vicious and cheap construction."[1] Derleth went on to say, "Rather
than try to tell outsiders that San Francisco was visited by a conflagration I
believe that it will do San Francisco and California in general more lasting
good to admit that there was an earthquake, and that with honest and intelligent construction and the avoidance of weak locations for important structures, our losses within the earthquake belt would not have been so great."[2]

If the principal interest of a contractor is profit, not safety, and the principal motivation of the urban planner is to maximize the number of dwellings
for a given outlay, we are heading for an urban society that must accept the
occasional collapse of a city or two. But where disasters occur because building codes are ignored, Nick Ambraseys has pointed out that these disasters
are not acts of God as much as acts of criminal negligence.

Yet contractors and urban planners are not seismologists, and they are
largely unaware of the risks they may be taking. So who must take the blame
for the disasters of the future? With some dismay we conclude that the world's
seismologists must take some of this responsibility. We are aware of the
planet's seismic legacy, and its inescapable future. Earthquake engineers have
the fix, but seismologists are the ones who spend their lives examining the
minutiae of earthquakes, the esoteric details of seismic waves, and historical
damage from earthquakes. Do seismologists spend too much time talking to
each other, and insufficient time talking to the urban planners of our future
world? The question is, at the very least, food for thought.

To return for a moment to the theme of this chapter, we live in an age of
construction: an age in which we are doubling the human inventory of dwellings in less than the span of a human lifetime. In 50 years we not only double
the building stock, we will replace many older buildings in our great cities
with newer structures. Why should they not be safer structures? These, after
all, are the buildings that will kill our descendants unless we act wisely. Manda-
tory earthquake resistance on all present and future construction would
yield a much safer world.

The argument eventually becomes one of economy. It has been estimated
that the cost of a building that can resist earthquake damage is typically at
most 10 percent more—and very often even less—than one that will collapse.
Simple measures cost a little less, and infallible measures cost a lot more. The
difference is between a ruined building from which its survivors can safely
escape (e.g., a low-cost dwelling), and an unblemished building that can function through an earthquake and require no post-earthquake repairs (e.g., a

hospital). Retrofitting a building may cost more than 40 percent of the original construction cost. The decision is a no-brainer if we are motivated to save lives, but it becomes much more difficult if the decision is based purely on economical arguments. We estimate that less than 1 percent of our world building stock will collapse in earthquakes, yet the introduction of earthquake resistance worldwide will cost as much as 10 percent more. This extra expense will be considered a poor investment in a world economy driven to take the lowest bid, or to maximize internal volume for minimal outlay. Earthquake disasters are therefore most certainly in our future. The cost of such disasters can moreover be enormous. For an individual building owner, the low-bid approach might make sense, especially given the likelihood that government assistance will be available if a large earthquake does strike. For society as a whole, the proverbial ounce of prevention is truly worth the proverbial pound of cure.

The Homes of Tomorrow

Problems are often easy enough to identify. Solutions are another matter.

A small handful of places around the globe have had the will and the resources to tackle their earthquake problem in earnest. Vulnerabilities remain in even California and Japan, but decades of commitment to stringent building codes have made a real difference—especially in saving lives. When the M6.7 Northridge earthquake struck the heart of the densely populated San Fernando Valley in 1994, the disaster was enormously expensive but claimed about as many lives as did the M7.6 Owens Valley earthquake, which struck the sparsely populated eastern Sierra region over a century earlier.

Within the United States the battle is still not entirely won, however. Some California lawmakers continue to challenge the stringent and expensive provisions of the Field Act, which ensures that California's public schools (although not its public universities) are built to some of the most stringent standards on the planet. In the central United States, seismologist Seth Stein has argued that it is not cost-effective or generally sensible that the midcontinent should "build for California earthquakes."[3] If one is paying a 2–10 percent surcharge to protect the 1 percent of buildings that would have otherwise collapsed, a dispassionate dollars-and-cents consideration might lead one to conclude that the costs are unacceptably high. A similar dispassionate dollars-and-cents consideration might lead one to conclude that it is senseless to pay for medical insurance, or car insurance, or homeowner's insurance.

Most people accept the costs of insurance, not as an investment that will "pay off," in the sense of returning more than one has paid in, but rather as insurance against foreseeable bad luck. Many—perhaps most—seismologists have the same view of earthquake risk mitigation. It is the price that we pay, as individuals and as a society, to protect ourselves—to protect our children—against an eventuality that might be unlikely but is still entirely possible. And the business of hazard mapping, such as that undertaken in recent years in the United States by the U.S. Geological Survey and the California Geological Survey, allows us to quantify with some precision the meaning of the term "entirely possible." In particular, such efforts allow scientists to identify those areas where damaging shaking might reasonably be expected to occur during the lifetime of a structure. Within the United States, such regions include not only California, but also the Pacific Northwest, the New Madrid (central U.S.) region, and coastal South Carolina. Damaging earthquakes are also possible, although less probable in other areas, including other western states as well as the Saint Lawrence Seaway. (One must also remember that damaging earthquakes are *possible* virtually everywhere on earth.)

To address remaining earthquake risk issues in places like Japan and California, the solutions are relatively straightforward, for private homeowners especially. Advice abounds, on the Web and elsewhere, on nonstructural risk mitigation steps: strapping one's water heater to the wall, that sort of thing. Local contractors, in some cases earthquake retrofit specialists, provide ready expertise on structural matters. Residents of multiple-unit housing, either condos or apartments, face a bigger challenge. In an older masonry building, or one with tuck-under parking, substantial retrofitting may be required to make a structure earthquake safe, yet such changes are beyond the power of the individual to effect.

In parts of the United States, as well as other industrialized countries, where earthquakes are infrequent but possible, expensive retrofitting might not be warranted, but increasingly stringent building codes should provide at least a measure of protection to those who will live in buildings now under construction. Very tall buildings have, moreover, been earthquake resistant by default for some time: such structures have to be designed to withstand enormous wind stresses.

The solution is far more difficult, of course, in developing nations—precisely those countries in which the population boom is occurring. With critical unmet needs in industrialized nations, and seemingly overwhelming unmet needs in developing nations, the cost of keeping the world's poor safe from earthquakes can easily appear prohibitive. Clearly, one must worry about the absolute necessities of life—food, water, shelter, basic medical care— before one worries about risk mitigation.

The problem—or, rather, one of the problems—with this line of reasoning is the title for this chapter. The Age of Construction is already under way: we have a unique opportunity to pay the modest surcharge now to protect the present *and future* inhabitants of the dwellings now under construction. The cost of retrofitting existing structures far exceeds that of building them right in the first place. Some risk mitigation efforts are already under way. Seismologist Brian Tucker launched GeoHazards, Incorporated, a small, nonprofit organization that connects the rubber to the road, raising funds for projects such as retrofitting school buildings in several developing nations. Global risk mitigation projects are also supported by a number of agencies, such as the World Bank.

Efforts such as these are good first steps, but only that. Without a redoubled commitment to earthquake risk mitigation, the future of earthquake impact worldwide is as grimly predictable as are the earthquake statistics on which they are based. Eventually a large earthquake will strike in the heart of an urban center, and claim a million lives. Joseph Stalin observed that "one death is a tragedy; a million deaths is a statistic." That he might have been more or less correct in his assessment of how humans view fatality on a large scale is beside the point: this is clearly not a person whose words one strives to live by. Perhaps a disaster on this scale is already inevitable, given existing structures and conditions in cities like New Delhi and Kathmandu. Nevertheless, *Homo sapiens* stands poised on the cusp of a transition of historic proportions: quite possibly the largest and last baby boom and building boom this planet will ever witness. We will not get a second chance to get it right the first time.

Notes and Suggested Readings

Chapter 1

NOTES

1. N. S. Shaler, "The Stability of the Earth," *Scribner's Magazine* (March 1887): 259.

2. Aristotle. Early theories of earthquakes, including Aristotle's, are described in a first-rate compilation available from the Historical Earthquake Theories Web site, http://www.univie.ac.at/Wissenschaftstheorie/heat

3. Robert Hooke, *Micrographia.* London: Royal Society of London, 1665; also in *The Posthumous Works of Robert Hooke,* ed. R. Waller (London: Samuel Smith and Benjamin Walford, 1705).

4. Charles Davison, *The Founders of Seismology* (Cambridge: Cambridge University Press, 1927), 185.

5. Benjamin F. Howell, Jr., *An Introduction to Seismological Research* (Cambridge: Cambridge University Press, 1990), 60.

6. Steven Jay Gould, *The Flamingo's Smile: Reflections in Natural History* (New York: W.W. Norton, 1985), 413.

SUGGESTED READINGS

Ben-Avraham, Zvi, and Susan Hough. "Promised Land." *Natural History* (October 2003): 44–49.

Bolt, Bruce A. *Earthquakes,* 5th ed. New York: W. H. Freeman, 2003.

Historical Earthquake Theories Web site. http://www.univie.ac.Wissenschafts theorie/heat

Hough, Susan Elizabeth. *Earthshaking Science: What We Know (and Don't Know) About Earthquakes.* Princeton, N.J.: Princeton University Press, 2002.

Kious, W. J., and Robert I. Tilling. *This Dynamic Earth.* Washington, D.C: U.S. Geological Survey, 1994/1996. See also http://pubs.usgs.gov/publications/text/ dynamic.html

Oeser, Erhard. Historical Earthquake Theories Web site. http://www.univie.ac.at/ Wissenschaftstheorie/heat/

Sieh, Kerry, and Simon LeVay. *The Earth in Turmoil: Earthquakes, Volcanoes, and Their Impact on Humankind.* New York: W. H. Freeman, 1998.

Yeats, Robert S. *Living with Earthquakes in California: A Survivor's Guide.* Corvallis: Oregon State University Press, 2001.

Chapter 2

NOTES

1. Rufus Steele, *The City That Is: The Story of the Rebuilding of San Francisco in Three Years* (San Francisco: A. M. Robertson, 1909), 32.

2. Amos Nur, "The Collapse of Ancient Societies by Great Earthquakes," In *Natural Catastrophes During Bronze Age Civilisations,* Benny J. Peiser, Trevor Palmer, and Mark E. Bailey, eds. (Oxford: Archaeopress, 1998).

SUGGESTED READINGS

De Boer, J. Z., and D. T. Sanders. *Earthquakes in Human History: The Far-Reaching Effects of Seismic Disruptions.* Princeton, N.J.: Princeton University Press, 2005.

Fields, Nic. *Troy c. 1700–1250 BC.* Oxford: Osprey, 2004.

Nur, Amos, and Ron Hagai. "Armageddon's Earthquakes." *International Geology Review 39* (1997): 532–541.

Chapter 3

NOTES

1. Charles Davison, *The Founders of Seismology* (Cambridge: Cambridge University Press, 1927), 1.

2. Rev. Charles Davy's account of the 1755 temblor, from *The World's Story: A History of the World in Story, Song, and Art,* vol. 5, ed. Eva March Tappan. Boston: Houghton Mifflin, 1914. See http://www.fordham.edu/halsall/mods/1755lisbonquake.html

3. Ibid.

4. Ibid.

5. Ibid.

6. Ibid.

7. Ibid.

8. Ibid.

9. Ibid.

10. Immanuel Kant, "What Is Enlightenment?" In *Foundation of the Metaphysics of Morals and What Is Enlightenment?,* trans. Lewis White Beck, 2nd rev. ed. (London: Collier Macmillan, 1990), 85.

11. Voltaire, "Optimism," in his *Philosophical Dictionary* (Works 12: 82–83), 1764.

12. Colin Brown, *Christianity and Western Thought,* vol. 1 (Downers Grove, Ill.: Intervarsity Press, 1990), 293.

13. Immanuel Kant, *Critique of Pure Reason,* trans. Norman Kemp Smith. An electronic edition available at http://www.arts.cuhk.edu.hk/Philosophy/Kant/cpr

14. John Wesley, "The Cause and Cure of Earthquakes," sermon (1750). http://www.segen.com/wesley/sermon04.html

15. Davison, *Founders of Seismology,* 17.

16. Ibid.

17. Russell R. Dynes, *The Dialogue between Voltaire and Rousseau on the Lisbon Earthquake: The Emergence of a Social Science View* (abstract), 4th European Conference of Sociology, Aug. 18–21, 1999 (Amsterdam: Vrije Universiteit, 2000).

SUGGESTED READINGS

Kendrick, Thomas D. *The Lisbon Earthquake.* Philadelphia: J. B. Lippincott, 1955.

Mitchell, John. "Conjectures Concerning the Cause, and Observations upon the Phaenomena, of Earthquakes." *Philosophical Transactions of the Royal Society of London* 2 (1760): 566–574.

Tappan, Eva March, ed. *The World's Story: A History of the World in Story, Song, and Art,* vol. 5. Boston: Houghton Mifflin, 1914.

Chapter 4

NOTES

1. James Lal Penick, Jr., *The New Madrid Earthquakes,* rev. ed. (Columbia: University of Missouri Press, 1994), 124.

2. John Sugden, *Tecumseh: A Life* (New York: Henry Holt, 1998), 255.

3. "The New Madrid extract" is among the accounts compiled by Ronald Street in *The Historical Seismicity of the Central United States: 1811–1928, Final Report,* appendix A (Washington, D.C.: U.S. Geological Survey, 1984). See http://www.ceri.memphis.edu/compendium

4. http://www.ceri.memphis.edu/compendium

5. http://www.ceri.memphis.edu/compendium

6. Henry Marie Brackenridge, *Views of Louisiana; Together with a Journal of a Voyage up the Missouri River, in 1811* (Pittsburgh, Pa.: Cramer, Spear, and Eichbaum, 1814), 212.

7. John Bradbury, *Travels in the Interior of America in the Years 1809, 1810, and 1811, Including a Description of Upper Louisiana.* (London: Sheewood, Neely, and Jones, 1817), 201.

8. Account by Eliza Bryan, available at http://www.ceri.memphis.edu/compendium

9. Account from Frankfort, Kentucky, available at http://www.ceri.memphis.edu/compendium

10. Account by John Bradbury, available at http://www.ceri.memphis.edu/compendium

11. Account by Firmin La Roche, available at http://www.ceri.memphis.edu/compendium

12. Account by William Leigh Pierce, available at http://www.ceri.memphis.edu/compendium

13. Letter from Eliza Bryan to Lorenzo Dow, available at http://www.ceri.memphis.edu/compendium

14. *Lexington Reporter,* available at http://www.ceri.memphis.edu/compendium

15. Account by James McBride, available at http://www.ceri.memphis.edu/compendium

16. Ibid.

17. Account by John Hardeman Walker, "Legend of the Memorable Earthquake of 1811," in Samuel Cummings, *The Western Pilot; Containing Charts of the Ohio River and the Mississippi, from the Mouth of the Missouri to the Gulf of Mexico; Accompanied with Directions for Navigating the Same, and a Gazetteer; or Description of the Towns on Their Banks, Tributary Streams, Etc., Also, a Variety of Matter Interesting to Travelers, and All Concerned in the Navigation of Those Rivers; with a Table of Distances from Town to Town on All the Above Rivers* (Cincinnati, Ohio: George Conclin, 1847), 140.

18. Ibid., 139.

19. Ibid.

20. Ibid., 140.

21. Ibid.

22. Ibid., 142.

23. Account by George Roddell, available at http://www.ceri.memphis.edu/compendium

24. Letter from Eliza Bryan to Lorenzo Dow, available at http://www.ceri.memphis.edu/compendium

25. Account by Robert McCoy, available at http://www.ceri.memphis.edu/compendium

26. Account by Col. John Shaw, available at http://www.ceri.memphis.edu/compendium

27. Account by W. Shaler, available at http://www.ceri.memphis.edu/compendium

28. Account by Mathias Speed, available at http://www.ceri.memphis.edu/compendium

29. Ibid.

30. Account by Vincent Nolte, available at http://www.ceri.memphis.edu/compendium

31. Ibid.

32. Account by Mathias Speed, available at http://www.ceri.memphis.edu/compendium

33. Account by W. Shaler, available at http://www.ceri.memphis.edu/compendium

34. Myron L. Fuller, *The New Madrid Earthquake* (Washington, D.C.: Government Printing Office, 1912), 75.

35. Samuel Mitchell, "A Detailed Narrative of the Earthquakes Which Occurred on the 16th Day of December, 1811." *Transactions of the Literary and Philosophical Society of New York 1* (1815): 281–307.

36. Amos Andrew Parker, *Parker's Trip to the West and Texas* (Concord, N.H.: White and Fisher, 1834), 257.

37. James Finley, *Autobiography of Rev. James B. Finley*, ed. W. P. Strickland (Cincinnati, Ohio: Methodist Book Concern, 1853), 238.

38. Walter Brownlow Posey, "The Earthquake of 1811 and Its Influence on Evangelistic Methods in the Churches of the Old South," *Tennessee Historical Magazine 1–2* (1931): 110.

39. Benjamin Casseday, *The History of Louisville from Its Earliest Settlement Till the Year 1852* (Louisville, Ky.: Hull and Brothers, 1852), 125–126.

40. Ibid.

SUGGESTED READINGS

Drake, Daniel. *Natural and Statistical View, or Picture of Cincinnati and the Miami County, Illustrated by Maps*. Cincinnati, Ohio: Looker and Wallace, 1815. http://pasadena.wr.usgs.gov/office/hough/Drake.html

McMurtrie, H., M.D. *Sketches of Louisville and Its Environs, Including, Among a Great Variety of Miscellaneous Matter, a Florula Louisvillensis; or, a Catalogue of Nearly 400 Genera and 600 Species of Plants, That Grow in the Vicinity of the Town, Exhibiting Their Generic, Specific, and Vulgar English Names*. Louisville, Ky.: S. Penn, Jr., 1839. http://pasadena.wr.usgs.gov/office/hough/Brooks.html

Penick, James Lal, Jr. *The New Madrid Earthquakes*, rev. ed. Columbia: University of Missouri Press, 1994.

Shaler, N. S. *Story of Our Continent: A Reader in the Geography and Geology of North America*. Boston: Ginn, 1899.

Chapter 5

NOTES

1. Charles Davison, *A Study of Recent Earthquakes* (London: Walter Scott, 1905), 10.

2. Ibid., 11.

3. Ibid.

4. Ibid.

5. Athanasius Kircher, "Earthquake at Calabria, in the Year 1638." http://www
.circoloalabrese.org/library/history/earthquake1638.asp

6. Ibid.

7. Ibid.

8. Ibid.

9. Ibid.

10. Charles Davison, *The Founders of Seismology* (Cambridge: Cambridge University Press, 1927), 65.

11. Ibid., 79.

12. Davison, *Study of Recent Earthquakes*, 266–267.

13. Richard D. Oldham, "Report on the Great Earthquake of 12 June 1897," *Memoirs of the Geological Society of India* 29 (Calcutta: Geological Survey of India, 1899), 379.

14. Davison, *Study of Recent Earthquakes*, 318.

15. Sir Harold Jeffreys, "An Interview with Henry Spall," U.S. Geological Survey, Reston, Va. Available at http://wwwneic.cr.usgs.gov/neis/seismology/people/int-jeffreys.html

Chapter 6

NOTES

1. Clarence Edward Dutton, "The Charleston Earthquake of August 31, 1886." In *Ninth Annual Report of the United States Geological Survey, 1887–88* (Washington, D.C.: Government Printing Office, 1889), 230. Dutton's authoritative report and compilation of accounts is the source for most of the accounts in this chapter.

2. Ibid., 231.

3. Ibid., 272.

4. Ibid.

5. Ibid., 212.

6. Ibid.

7. Ibid., 213.

8. Ibid.

9. Ibid., 214.

10. Ibid., 221.

11. Paul Pinckney, letter to *The San Francisco Chronicle*, May 6, 1906.

12. Dutton, "Charleston Earthquake," plate XXVI.

13. J. G. Armbruster and S. E. Hough, "1886 Earthquake Studies," *Eos: Transactions of the American Geophysical Union* 82 (August 2001): 349.

14. Dutton, "Charleston Earthquake," 388.

15. Ibid., 389.

16. Ibid.

17. Ibid., 409.

18. Ibid., 211.

Chapter 7

NOTES

Many of the accounts in this chapter have been republished in Malcolm E. Barker, *Three Fearful Days: San Francisco Memoirs of the 1906 Earthquake and Fire* (San Francisco: Londonborn Publications, 1998).

1. Fritz Foote, unpublished photo album, entry date April 18, 1900.

2. Ibid.

3. Mayor E. E. Schmitz's infamous "shoot to kill" order was issued on April 18, 1906. See http://www.sfmuseum.org/1906.2/killproc.html

4. Foote, photo album, entry date April 19, 1906.

5. Jack London, "The Story of San Francisco," *Colliers, The National Weekly*, May 5, 1906.

6. John C. Worth and Marcia M. Worth, unpublished letter, dated May 13, 1906.

7. French Strother, "The Rebound of San Francisco," *Harper's Magazine*, May 1906.

8. Gladys Hansen and Emmett Condon, *Denial of Disaster* (San Francisco: Cameron, 1989), 111.

9. Ibid., 108.

10. John Caspar Branner, "Earthquakes and Structural Engineering," *Bulletin of the Seismological Society of America 3* (1913): 2.

11. Ibid., 3.

12. Hansen and Condon, *Denial of Disaster*, 107.

13. Grove Karl Gilbert, *The Investigation of the California Earthquakes* (San Francisco, A. M. Robertson, 1907), 242.

14. Strother, "The Rebound of San Francisco."

15. William James, "Some Mental Effects of the Earthquake," *Youth's Companion*, June 7, 1906. Reprinted in Barker, *Three Fearful Days*, 297.

16. Ibid.

17. Ibid.

18. Pauline Jacobson, "How It Feels to Be a Refugee and Have Nothing in the World," *San Francisco Bulletin*, April 29, 1906. Reprinted in Barker, *Three Fearful Days*, 283.

19. Ibid.

20. Ibid.

21. http://www.sfmuseum.org

22. Winifred Black Bonfils, writing as Annie Laurie, "Annie Laurie Tells of the Spectral City," *San Francisco Examiner*, April 22, 1906. Reprinted in Barker, *Three Fearful Days*, 219.

23. Mary Austin, "The Temblor: A Personal Narration," in *The California Earthquake of 1906*, ed. David Starr Jordan (San Francisco: A. M. Robertson, 1907), 351–352.

24. Hansen and Condon, *Denial of Disaster*, 116.

25. Rufus Steele, *The City That Is: The Story of the Rebuilding of San Francisco in Three Years* (San Francisco: A. M. Robertson, 1909), 83–84.

26. Edward M. Lind, "How They Saved Hotaling's Whiskey," in Barker, *Three Fearful Days*, 188.

27. Ibid., 199.

28. Ibid.

29. Edwin Emerson, "San Francisco at Play," *Sunset Magazine*, October 1906. Reprinted in Barker, *Three Fearful Days*, 309.

30. Ibid.

31. Steele, *The City That Is*, 19.

32. Ibid., 96.

33. Ibid.

34. Ibid., 19.

35. William Prescott, "Circumstances Surrounding the Preparation and Suppression of a Report on the 1868 California Earthquake," *Bulletin of the Seismological Society of America 72* (1982): 2391.

36. Ibid., 2392.

37. Ibid., 2391.

38. Ibid., 2392.

39. Hansen and Condon, *Denial of Disaster*, 121.

40. *Earthquakes and Building Construction: A Review of Authoritative Engineering Data and Records of Experience* (Los Angeles: Clay Products Institute of California, 1929), 14.

41. Charles Derleth, Jr., "The Destructive Extent of the California Earthquake of 1906; Its Effect upon Structures and Structural Materials, within the Earthquake Belt," in Jordan, *California Earthquake*, 100.

42. Harry Fielding Reid, in Andrew C. Lawson, *The California Earthquake of April 18, 1906: Report of the State Earthquake Investigation Commission. Volume II: The Mechanics of the Earthquake, by Harry Fielding Reid* (Washington D.C.: Carnegie Institution, 1908), 27.

43. Charles Davison, *The Founders of Seismology* (Cambridge: Cambridge University Press, 1927), 152–153.

44. Ibid., 152.

45. G. K. Gilbert, "The Earthquake as a Natural Phenomenon," in *The San Francisco Earthquake and Fire of April 18, 1906, and Their Effects on Structures and Structural Materials*, U.S. Geological Survey Bulletin 324 (1907), 1–13.

46. Lawson, *California Earthquake*, 27.

47. Ibid., 31–32.

48. Ibid., 32.

49. Ibid. (account by Fairbanks), 38.

50. Ibid.

51. Ibid., 40.

52. Ibid.

53. Ibid., 41.

54. Ibid.

55. Ibid.

56. Ibid.

57. Ibid.

58. Ibid.

59. Ibid., 42.

60. Ibid., 43.

61. Ibid.

62. Ibid.

63. Ibid., 44.

64. Ibid., 45.

65. Ibid., 46.

66. Ibid., 47.

67. Ibid. (account by Lawson), 48.

68. Ibid., 52.

69. Ibid.

70. Ibid., 42.

SUGGESTED READINGS

Collier, Michael. *A Land in Motion: California's San Andreas Fault*. Berkeley: University of California Press, 1999.

Hough, Susan Elizabeth. *Finding Fault in California: An Earthquake Tourist's Guide*. Missoula, Mont.: Mountain Press, 2004.

http://www.sfmuseum.org/1906/06.html. A treasure trove of information, accounts, etc.

McClellan, Rolander Guy. *Golden State: A History of the Region West of the Rocky Mountains, Embracing California, Oregon, Nevada, Utah, Arizona, Idaho, Washington Territory, British Columbia, and Alaska, from the Earliest Period to the Present Time*. Philadelphia: William Flint, 1872.

Wallace, Robert, ed. *The San Andreas Fault System*. U.S. Geological Survey Professional Paper 1515. Washington, D.C.: U.S. Geological Survey, 1990.

Chapter 8

NOTES

1. Otis Manchester Poole, *The Death of Old Yokohama in the Great Japanese*

Earthquake of 1923 (London: George Allen and Unwin, 1968), 31. This publication provides most of the firsthand accounts in this chapter.

2. Ibid., 76.

3. *Earthquakes and Building Construction: A Review of Authoritative Engineering Data and Records of Experience* (Los Angeles: Clay Products Institute of California, 1929), 31.

4. Radio telegram from Tokyo, sent by Baron Okura, available at http://nisee .berkeley.edu/kanto/kanto.html

5. Frank Lloyd Wright, see http://nisee.berkeley.edu/kanto/kanto.html

6. Ibid.

7. Ibid.

8. K. Sawada and Charles A. Beard, "Reconstruction in Tokyo" (March 1925): 268.

9. Poole, *Death of Old Yokohama,* 20.

10. Ibid., 132.

11. Ibid., 130.

12. Ibid.

13. William Elliot Griffis, "The Empire of the Risen Sun," *National Geographic* (October 1923): 441.

14. Ibid., 415.

15. "Speaking of Pictures . . . These Show That Horror Is Not New to the Japanese," *Life* (August 16, 1943): 10–11.

16. Inside cover of tourist booklet, "Yokohama," printed by Ohkawa Printing Company, Yokohama, Japan (publication date unknown).

SUGGESTED READINGS

Poole, Otis Manchester. *The Death of Old Yokohama in the Great Japanese Earthquake of 1923.* London: George Allen and Unwin, 1968. This book, although difficult to find, provides a remarkable firsthand account of the aftermath of the Kanto earthquake.

Seidensticker, Edward. *Tokyo Rising: The City Since the Great Earthquake.* Cambridge, Mass.: Harvard University Press, 1991.

Chapter 9

NOTES

1. Louis V. Housel, "An Earthquake Experience," *Scribner's Monthly* 15, no. 5 (March 1878): 662. This remarkable account provides many of the firsthand descriptions of the 1867 earthquake and tsunami in this chapter.

2. Ibid., 666.

3. Ibid.

4. Ibid., 667.

5. Ibid.

6. Ibid., 668.

7. Ibid., 671.

8. Ibid., 671–672.

9. Ibid., 672.

Chapter 11

NOTES

1. Mike Davis, *Ecology of Fear* (New York: Henry Holt, 1998).

2. Tom O'Keefe, quoted by Associated Press, October 30, 2003.

3. Anonymous quote to author, 2004.

4. James Dolan, direct quote to author, 2004.

5. Rolander Guy McClellan, *Golden State: A History of the Region West of the Rocky Mountains, Embracing California, Oregon, Nevada, Utah, Arizona, Idaho, Washington Territory, British Columbia, and Alaska, from the Earliest Period to the Present Time* (Philadelphia: William Flint, 1872), 222.

6. A. H. Godbey, *Great Disasters and Horrors in the World's History* (St. Louis, Mo.: Imperial Publishing, 1890), 556.

7. Mona and Ivan Hecksher, "Southern California Fires: A Whole Town Is Gone," quoted by John Balzar, *Los Balzar* (October 31, 2003): A-1.

8. http://www.kfwb.com, November 2003.

SUGGESTED READING

Davis, Mike. *Ecology of Fear.* New York: Henry Holt, 1998.

Chapter 12

NOTES

1. A. Owen Aldridge, *Voltaire and the Century of Light* (Princeton, N.J.: Princeton University Press, 1975), 302.

2. Paul Pinckney, letter to the *San Francisco Chronicle*, May 6, 1906.

3. *San Francisco Real Estate Circular*, October 1868. See http://www.sfmuseum.org/hist1/1868.html

4. A. H. Godbey, *Great Disasters and Horrors in the World's History* (St. Louis, Mo.: Imperial Publishing, 1890), 535–536.

5. Ibid., 536.

Chapter 13

SUGGESTED READINGS

Chandler, Tertius. *Four Thousand Years of Urban Growth: An Historical Census.* Lewiston, N.Y.: Edwin Mellon Press, 1987.

Chase-Dunn, Christopher. "The System of World Cities: A.D. 800–1975," in *Urbanization in the World Economy,* ed. Michael Timberlake. New York: Academic Press, 1985.

Chen, Yong, ed. *The Great Tangshan Earthquake of 1976: An Anatomy of Disaster.* Oxford: Pergamon, 1988.

Dunbar, P. K., P. A. Lockridge, and L. S. Whiteside. *Catalog of Significant Earthquakes 2150 B.C. to 1991 A.D., Including Quantitative Casualties and Damage.* National Geophysical Data Center Report SE49. Washington, D.C.: National Geophysical Data Center, 1992.

Chapter 14

NOTES

1. Charles Derleth, Jr., "The Destructive Extent of the California Earthquake of 1906; Its Effect upon Structures and Structural Materials, Within the Earthquake Belt," in *The California Earthquake of 1906,* ed. David Starr Jordan (San Francisco: A. M. Robertson, 1907), 101.

2. Ibid., 101–102.

3. Seth Stein, "Should Memphis Build for California Earthquakes?" *EOS: Transactions of the American Geophysical Union* (May 13, 2003): 177.

SUGGESTED READINGS

Hansen, Gladys, and Emmet Condon. *Denial of Disaster.* San Francisco: Cameron, 1989.

Jordan, David Starr, ed. *The California Earthquake of 1906.* San Francisco: A. M. Robertson, 1907.

Lawson, Andrew. *The California Earthquake of April 18, 1906.* Washington, D.C.: Carnegie Institution, 1908.

McClellan, Rolander Guy. *Golden State: A History of the Region West of the Rocky Mountains, Embracing California, Oregon, Nevada, Utah, Arizona, Idaho, Washington Territory, British Columbia, and Alaska, from the Earliest Period to the Present Time.* Philadelphia: William Flint, 1872.

Index